PURE AND APPLIED MATHEMATICS

A Program of Monographs, Textbooks, and Lecture Notes

LECTURE NOTES
IN PURE AND APPLIED MATHEMATICS

1. *N. Jacobson,* Exceptional Lie Algebras
2. *L.-A. Lindahl* and *F. Poulsen,* Thin Sets in Harmonic Analysis
3. *I. Satake,* Classification Theory of Semi-Simple Algebraic Groups
4. *F. Hirzebruch, W. D. Neumann,* and *S. S. Koh,* Differentiable Manifolds and Quadratic Forms
5. *I. Chavel,* Riemannian Symmetric Spaces of Rank One
6. *R. B. Burckel,* Characterization of C(X) among Its Subalgebras
7. *B. R. McDonald, A. R. Magid,* and *K. C. Smith,* Ring Theory: Proceedings of the Oklahoma Conference
8. *Yum-Tong Siu,* Techniques of Extension of Analytic Objects
9. *S. R. Caradus, W. E. Pfaffenberger, and Bertram Yood,* Calkin Algebras and Algebras of Operators on Banach Spaces
10. *Emilio O. Roxin, Pan-Tai Liu,* and *Robert L. Sternberg,* Differential Games and Control Theory
11. *Morris Orzech* and *Charles Small,* The Brauer Group of Commutative Rings
12. *S. Thomeier*, Topology and its Applications
13. *Jorge M. López* and *Kenneth A. Ross*, Sidon Sets
14. *W. W. Comfort* and *S. Negrepontis*, Continuous Pseudometrics
15. *Kelly McKennon* and *Jack M. Robertson*, Locally Convex Spaces

Other volumes in preparation

LOCALLY CONVEX SPACES

LOCALLY CONVEX SPACES

Kelly McKennon / Jack M. Robertson

Washington State University
Pullman, Washington

MARCEL DEKKER, INC. New York and Basel

MARCEL DEKKER, INC.

270 Madison Avenue, New York, New York 10016

LIBRARY OF CONGRESS CATALOG CARD NUMBER: 75-40934

ISBN: 0-8247-6426-9

Current printing (last digit):
10 9 8 7 6 5 4 3 2 1

PRINTED IN THE UNITED STATES OF AMERICA

PREFACE

There are many different types of locally convex linear topological spaces, and many ways of constructing new locally convex spaces from old. While most of the connections which the various spaces have with one another and with the different constructions have been determined, they apparently have never been systematically presented or even all set down in one volume. This especially is true with regard to the counter-examples involved. While many are scattered through various textbooks and monographs, some are only to be found in the less accessible literature and then only presented in brief form.

Following a careful study of the theory of linear topological spaces which extended over a considerable length of time at Washington State University, there was a feeling that the examples studied had not been adequate for full appreciation of the general theory. During the summer of 1970 and 1971 an example-oriented seminar was conducted in which was collected a considerable share of the material presented in this paper. It seemed reasonable to offer the results of these efforts to the mathematical public.

In the first and second chapters are contained the definitions of the types of spaces and constructions to be considered in the paper. Chapter III show what types of spaces are preserved under what constructions, and Chapter IV shows what types of spaces are always other types of spaces.

If a result is of a positive nature, the authors have referred to a standard textbook for a proof. The choice of references has been made partially on the basis of accessibility to the public, but mainly on the basis of familiarity to the authors. Because of the frequency of these referrals the reference is given by letter (eg. [S] for Schaefer) rather than by number.

If a result is of a negative nature and requires a counter-example for its verification, that counter-example will be found in Chapter V. It is the counter-examples which are most difficult to find in the literature and, for this reason, the authors have included all those found necessary for this volume.

Several remarks on the terminology employed in this paper are in order. The term locally convex space will always be taken to be a Hausdorff locally convex linear topological space. The scalar field, usually denoted by the letter K, may be taken to be either the field R of real numbers or the field C of complex numbers. The term neighborhood as used here does not imply that the set in question is itself open. In general, other terminology is as in [S].

The authors are indebted for the support granted Dr. Robertson in the summer of 1970 and that granted Dr. McKennon in the summer of 1971 by Washington State University summer grants-in-aid.

Thanks are also due to Mr. Bert Carbaugh for his help in collecting and proofreading the matter of this work, and for some needed impetus, particularly in the early stages. Finally, we wish to thank Mrs. Helen Niven and Mrs. Pamela Terry for their patience and care in typing the manuscript.

Kelly McKennon
Jack Robertson
Washington State University

CONTENTS

LOCALLY CONVEX SPACES

I. CONSTRUCTIONS OF LOCALLY CONVEX SPACES

Of those standard methods for producing new locally convex spaces from old, the sequel contains those utilized in this paper.

A. Projective Constructions.

1. Locally Convex Projective Topology. Let $\{E_\alpha\}_{\alpha \in I}$ be a family of locally convex spaces and, for each $\alpha \in I$, let f_α be a linear operator from a fixed linear space E into E_α. If the family $\{f_\alpha\}_{\alpha \in I}$ separates points of E, then the coarsest topology on E for which each f_α is continuous is called a locally convex projective topology.

2. Locally Convex Product. If the space E in (A.1) is the product $\prod_{\alpha \in I} E_\alpha$, and if each f_α is the canonical projection of E onto E_α, then E, with the associated l.c. projective topology, is called a locally convex product.

3. Locally Convex Projective Limit. Let I be a directed set with ordering $\leq$, and let $\{E_\alpha\}_{\alpha \in I}$ be a family of locally convex spaces. For each $\alpha \in I$ and $\beta \in I$ for which $\alpha \leq \beta$, let $f_{\alpha\beta}$ be a continuous linear operator from E_β into E_α. Then, the set $\{x \in \prod_{\alpha \in I} E_\alpha : (\forall \alpha, \beta \in I: \alpha \leq \beta) x_\alpha = f_{\alpha\beta}(x_\beta)\}$ is a closed subspace of $\prod_{\alpha \in I} E_\alpha$. Under the induced product topology, it is called a locally convex projective limit and is written $\varprojlim f_{\alpha\beta} E_\beta$.

B. Inductive Constructions.

1. Locally Convex Inductive Topology. Let $\{E_\alpha\}_{\alpha \in I}$ be a family of locally convex spaces and, for each $\alpha \in I$, let f_α be a linear operator from E_α into a fixed linear space E. Let $\mathfrak{T}$ be the finest locally convex topology on E for which each f_α is continuous. If $\mathfrak{T}$

is Hausdorff, it will be called a locally convex inductive topology.

2. Locally Convex Direct Sum. If the space E in (B.1) is the algebraic direct sum $\bigoplus_{\alpha\in I} E_\alpha$, and if each f_α is the canonical injection of E_α into E, then the associated topology $\mathfrak{T}$ is a l.c. inductive topology and E, under $\mathfrak{T}$, is called a locally convex direct sum.

3. Locally Convex Quotient Space. Let E be a locally convex space and S a closed linear subspace. Then the finest topology for which the canonical map q of E onto the quotient space E/S is continuous, is a locally convex topology. Under this topology, E/S is called a locally convex quotient space.

4. Locally Convex Inductive Limit. Let I be a directed set with ordering $\leq$, and let $\{E_\alpha\}_{\alpha\in I}$ be a family of locally convex spaces. For each $\alpha\in I$ and $\beta\in I$ for which $\alpha\leq\beta$, let $h_{\beta\alpha}$ be a continuous linear operator from E_α into E_β. For each $\alpha\in I$, write i_α for the canonical imbedding of E_α into the l.c. direct sum $\bigoplus_{\alpha\in I} E_\alpha$. Write H for the linear subspace of $\bigoplus_{\alpha\in I} E_\alpha$ generated by the set $\{i_\alpha(x) - i_\beta \circ f_{\beta\alpha}(x)\colon x\in E_\alpha,\ \alpha, \beta\in I,\ \alpha\leq\beta\}$. If H is closed, then the l.c. quotient $\bigoplus_{\alpha\in I} E_\alpha/H$ is called a locally convex inductive limit and is written $\varinjlim h_{\beta\alpha} E_\alpha$.

Note: H is not, in general, closed. See the appendix.

5. Locally Convex Inductive Limit of Subspaces. Let the notation here be as in (B.4). Note that, for $x, y\in \bigoplus_{\alpha\in I} E_\alpha$, $x = y \bmod (H)$ iff

$$\sum_{\alpha\leq\beta} f_{\beta\alpha}(y_\alpha) = \sum_{\alpha\leq\beta} f_{\beta\alpha}(x_\alpha)$$

whenever $\beta\in I$ is such that $\gamma\leq\beta$ for all $\gamma\in I$ for which $y_\gamma \neq 0$ or $x_\gamma \neq 0$. Suppose that $E_\alpha \subset E_\beta$ for all $\alpha, \beta\in I$ for which $\alpha\leq\beta$, and suppose that $f_{\beta\alpha}$ is the identity mapping of E_α into E_β. Let $E = \bigcup_{\alpha\in I} E_\alpha$ and define $\psi| \bigoplus_{\alpha\in I} E_\alpha \to E$ by letting $\psi(x) = \sum_{\alpha\in I} x_\alpha$. Then H is the kernel of ψ and so

$\lim_{\rightarrow} h_{\beta\alpha} E_\alpha$ is isomorphic to E , if E bears the finest locally convex topology of which the restriction to each E_α is coarser than the topology on E_α . The space E is called an inductive limit of subspaces.

6. Strict Inductive Limit. Let $\{E_n\}$ be a sequence of locally convex spaces, each E_n contained in some fixed linear space E . suppose that $E = \bigcup_{n=1} E_n$ and that, for all natural numbers $n < m$, E_n is a closed subspace (topological as well as algebraic) of E_m . Then E, bearing the finest topology of which the restriction to any E_n is coarser than that on E_n , is said to be a <u>strict inductive limit</u> . This is a special case of (B.5).

II. TYPES OF LOCALLY CONVEX SPACES

In this chapter are listed and described those types of locally convex spaces to be considered in the sequel. By E will be meant a fixed, but arbitrary, locally convex space with topology $\mathfrak{T}$. The linear space E' consisting of the continuous linear functionals on E may bear various topologies: the weak topology $\sigma(E',E)$ is the topology of pointwise convergence on E ; the Mackey topology $\tau(E', E)$ (Arens topology $\varkappa(E', E)$) is the topology of uniform convergence on convex, circled, weakly compact (compact) subsets of E ; the strong topology $\beta(E', E)$ is the topology of uniform convergence on bounded subsets of E .

A. Polar Spaces.

1. If any of the following pairwise-equivalent conditions hold, then E is said to be a Mackey space:
   (i) $\mathfrak{T}$ is the finest locally convex topology on E for which E' is the dual of E ;
   (ii) each convex, circled, $\sigma(E', E)$-compact subset of E' is equicontinuous;
   (iii) for each $\mathfrak{T}$-neighborhood V of zero in E , there exists a convex, circled, $\sigma(E', E)$-compact subset A of E' for which $A^{\circ} \equiv \{ x \in E: |f(x)| \leq 1 \ (\forall f \in A) \}$ is a subset of V .

[S p. 142] .

2. A circled, convex, absorbent, closed subset of E is said to be a barrel. If any of the following pairwise equivalent conditions hold, then E is said to be infrabarreled:
   (i) each barrel in E which absorbs bounded sets is a neighborhood of zero;
   (ii) each $\beta(E', E)$-bounded subset of E' is equicontinuous;
   (iii) for each $\mathfrak{T}$-neighborhood V of zero in E , there exists a $\beta(E', E)$-bounded subset A of E' for which $A^{\circ} \subset V$.

[S p. 142] .

3. If any of the following pairwise-equivalent conditions hold, then E is said to be barreled:

(i) each barrel is a neighborhood of zero in E ;

(ii) each $\sigma(E',E)$-bounded subset of E' is equicontinuous;

(iii) for each $\mathfrak{T}$-neighborhood V of zero in E , there exists a $\sigma(E',E)$-bounded subset A of E' for which $A^\circ \subset V$.[S IV.5.2]

4. That Mackey, infrabarreled, and barreled are analogous spaces is illustrated by the following scheme of polars:

Family F_j of subsets of E	Family H_j of subsets of E'
$F_1 \equiv$ closed, circled, convex neighborhoods of zero	$H_1 \equiv$ closed, circled, convex equicontinuous sets
$F_2 \equiv$ closed, circled, convex Mackey neighborhoods of zero	$H_2 \equiv \sigma(E',E)$-compact, circled convex sets
$F_3 \equiv$ barrels which absorb bounded sets	$H_3 \equiv \beta(E',E)$-bounded, closed, circled convex sets
$F_4 \equiv$ barrels	$H_4 \equiv \sigma(E',E)$-bounded, closed, circled, convex sets

For each j = 1,2,3,4, H_j is the family of polars of elements of F_j , and vice-versa. Furthermore,

$F_1 \subset F_2 \subset F_3 \subset F_4$ and $H_1 \subset H_2 \subset H_3 \subset H_4$.

Also, E is Mackey $\iff F_1 = F_2 \iff H_1 = H_2$; E is infrabarreled $\Leftrightarrow F_1 = F_3 \Leftrightarrow H_1 = H_3$; E is barreled $\Leftrightarrow F_1 = F_4 \Leftrightarrow H_1 = H_4$.

Thus, a barreled space is infrabarreled and an infrabarreled space is a Mackey space.
[S IV.5]

B. Spaces Distinguished by Bounded Sets.

The linear space consisting of all $\beta(E',E)$-continuous linear functionals on E' is denoted by E'' . Define the embedding $\eta|E \to E''$ by letting

$$\eta_x(f) \equiv f(x)$$

for all $x \in E$ and $f \in E'$. Then η is a linear space isomorphism of E into E'' .

The strong topology $\beta(E'',E')$ on E'' is the topology of uniform convergence on $\beta(E',E)$-bounded subsets of E'. It is not difficult to see that η^{-1} is $\beta(E'',E')$-$\mathfrak{T}$ continuous and that η is a homeomorphism if and only if E is infrabarreled.

The weak topology $\sigma(E'', E')$ is the topology of pointwise convergence on E'.

That η is $\mathfrak{T}$-$\sigma(E'',E')$ continuous is trivial. The topology $\sigma(E,E')$ for which η is a $\mathfrak{T}$-$\sigma(E'',E')$ homeomorphism is entitled the weak topology on E .

1. If each closed and bounded subset of E is complete, then E is said to be quasi-complete.

2. If each closed and bounded subset of E is $\sigma(E,E')$-compact, then E is said to be semi-reflexive. A necessary and sufficient condition for E to be semi-reflexive is for η to be surjective.
[S IV.5.5]

3. If a semi-reflexive space E is barreled (or infrabarreled) then it is said to be reflexive. A necessary and sufficient condition for E to be reflexive is for η to be an isomorphism of E onto E'' which is a $\mathfrak{T}$-$\beta(E'',E)$ homeomorphism.
[S. IV.5.6].

4. If each closed and bounded subset of E is compact, then E is said to be semi-Montel.

5. If a semi-Montel space is barreled (or infra-barreled), it is said to be Montel.

6. The following schemes show the relations between the various classes of bounded sets for different spaces:
(Notation: "$\hookrightarrow$" means "is a subset of", "$\Rightarrow$" means "implies", and circled numbers refer to footnotes regarding the non-trivial inclusions)

$\mathfrak{E} \equiv \{A \subset E:$ A is closed, circled, and convex$\}$;
$\mathfrak{E}_b \equiv \{A \in \mathfrak{E}:$ A is bounded $\}$;
$\mathfrak{E}_{\sigma b} \equiv \{A \in \mathfrak{E}:$ A is $\sigma(E,E')$-bounded$\}$;
$\mathfrak{E}_{\beta b} \equiv \{A \in \mathfrak{E}:$ A is $\beta(E,E')$-bounded$\}$;
$\mathfrak{E}_p \equiv \{A \in \mathfrak{E}:$ A is totally bounded (or precompact)$\}$;
$\mathfrak{E}_{\sigma p} \equiv \{A \in \mathfrak{E}:$ A is $\sigma(E,E')$- precompact$\}$;
$\mathfrak{E}_{\beta p} \equiv \{A \in \mathfrak{E}:$ A is $\beta(E,E')$-precompact$\}$;
$\mathfrak{E}_c \equiv \{A \in \mathfrak{E}_b:$ A is complete$\}$;
$\mathfrak{E}_{\sigma c} \equiv \{A \in \mathfrak{E}_b:$ A is $\sigma(E,E')$-complete$\}$;
$\mathfrak{E}_k \equiv \{A \in \mathfrak{E}:$ A is compact$\}$;
$\mathfrak{E}_{\sigma k} \equiv \{A \in \mathfrak{E}:$ A is $\sigma(E,E')$-compact$\}$;
$\mathfrak{E}_{\beta k} \equiv \{A \in \mathfrak{E}:$ A is $\beta(E,E')$-compact$\}$;

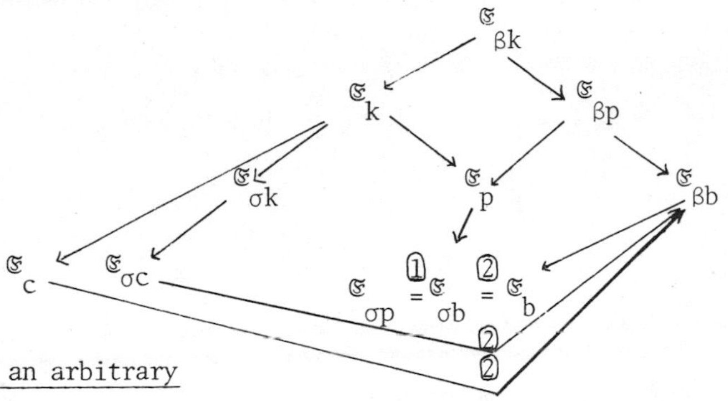

𝔗_{βk}

𝔗_k   𝔗_{βp}

𝔗_{σk}   𝔗_p

𝔗_{βb}

𝔗_c   𝔗_{σc}   ①𝔗 ②𝔗   𝔗_b
                𝔗_{σp} = 𝔗_{σb} =

②

②

E an arbitrary
l.c.s.

𝔗_{βk}

𝔗_k   𝔗_{βp}

𝔗_p

𝔗_{σk}

E a quasi-complete   𝔗_{σc} ⟶ 𝔗_c = 𝔗_{σp} = 𝔗_{σb} = 𝔗_b = 𝔗_{βb}
l.c.s.

𝔗_{βk}   𝔗_{βk}

𝔗_k   𝔗_p   𝔗_{βp}   𝔗_{βp}

𝔗_{σc} = 𝔗_c = 𝔗_{σk} = 𝔗_{σp} = 𝔗_{σb} =   𝔗_{σc} = 𝔗_c = 𝔗_{σk} = 𝔗_k = 𝔗_{σp} = 𝔗_p =

𝔗_b = 𝔗_{βb}   𝔗_{σb} = 𝔗_b = 𝔗_{βb}

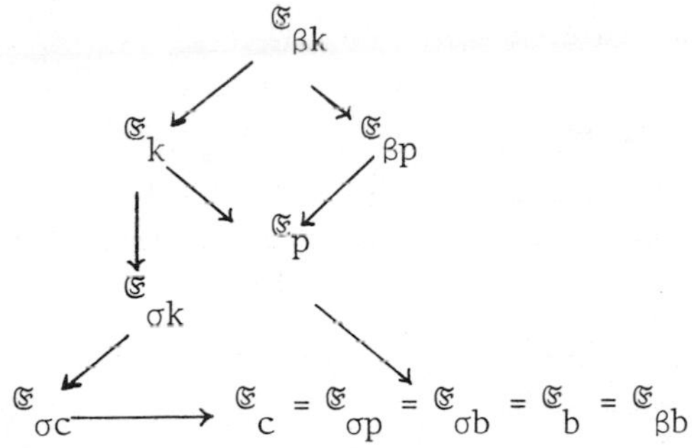

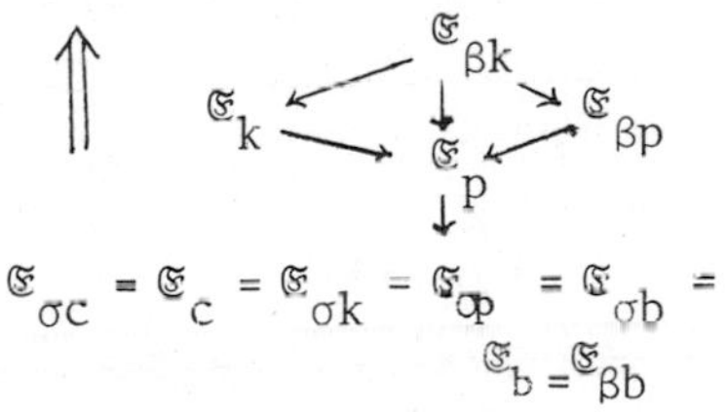

E a semi-reflexive l.c.s.   ⟸   E a semi-Montel space

𝔗_{βk} = 𝔗_k

𝔗_{bp} = 𝔗_p

𝔗_{σc} = 𝔗_c = 𝔗_{σk} = 𝔗_{σp} = 𝔗_{σb} = 𝔗_b = 𝔗_{βb}   𝔗_{σc} = 𝔗_c = 𝔗_{σk} = 𝔗_k = 𝔗_{βk} = 𝔗_{σp} = 𝔗_p = 𝔗_{βp} =

𝔗_{σb} = 𝔗_b = 𝔗_{βb}

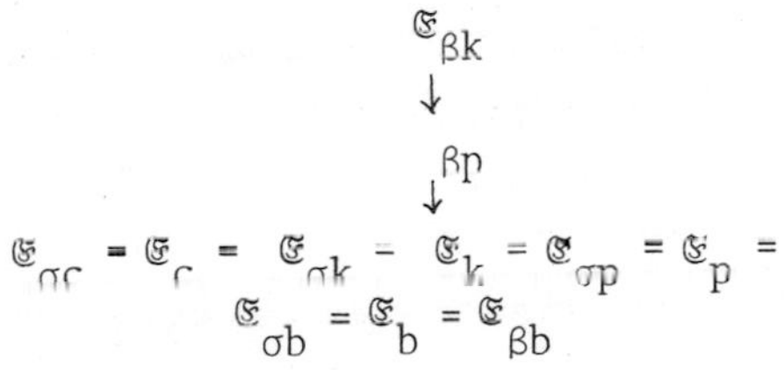

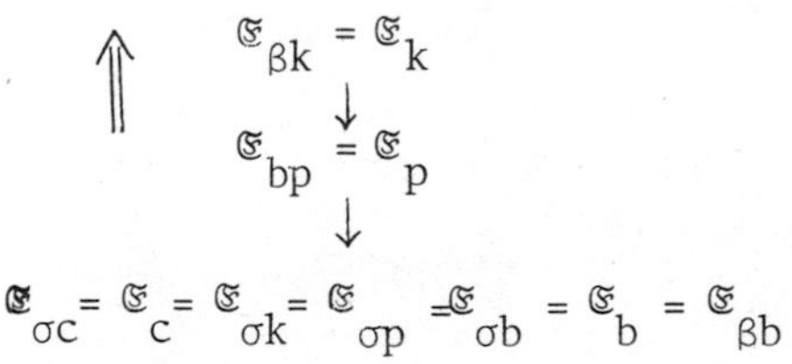

E a reflexive l.c.s.   E a Montel space

1. [R p.51]
2. [R p.67]

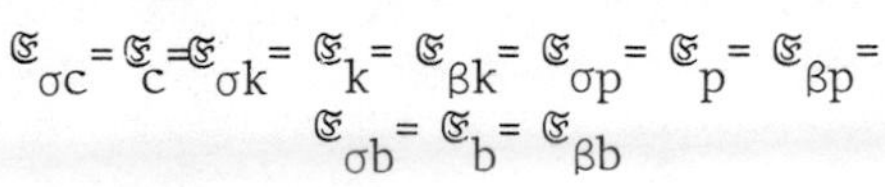

C. Spaces Distinguished by Continuity of Operators.

1. If any of the following pairwise equivalent conditions are satisfied, then E is said to be bornological:

(i) each linear operator on E into another locally convex space which preserves bounded sets (or compact sets) is continuous;

(ii) each circled, convex subset of E which absorbs bounded sets is a neighborhood of zero;

(iii) E is an inductive limit of normed spaces.

[S pp. 62-63], [R p. 83].

2. If any of the following pairwise equivalent conditions are satisfied, then E is said to be ultra-bornological:

(i) each linear operator on E into another locally convex space which preserves compact, circled, convex sets is continuous;

(ii) each circled, convex subset of E which absorbs circled, convex, compact sets is a neighborhood of zero;

(iii) E is an inductive limit of Banach spaces.

3. The relation which ultra-bornological spaces bear to bornological spaces is analogous to that which barreled spaces bear to infrabarreled spaces. Compare the statements (C.1.ii) and (C.2.ii) with the statements (A.2.i) and (A.3.i).

III. PRESERVATION OF PROPERTIES UNDER VARIOUS CONSTRUCTIONS

Suppose that $\{E_\alpha\}$ is a family of locally convex spaces, all E_α being of a common type (such as one of those defined in II). Suppose that a new locally convex space E is constructed from this family by one of the methods described in I. Is E of the same type as that of the E_α ? It is the purpose of this chapter to answer this question.

The table below lists various spaces in the first column and various constructions in the first row. A plus in a box means the type of space of that row is always preserved by the construction of that column; a minus means the property is not always preserved. To find a reference (in the case of a plus) or a counter-example (in the case of a minus), note the row and column numbers of the box, and see the material below:

1. <u>Mackey Spaces</u>.

(i) See (V.F.vii) and (V.F.viii).

(ii) [S], p. 138.

(iii) See (V.D.xiv) and (V.D.x).

(iv) Every locally convex space can be embedded in a product of Banach spaces. [R] p. 88. An embedding of (V.D.) in such a product produces a closed (V.D.xiii) non-Mackey (V.D.x) subspace.

(v) Suppose that E has a locally convex inductive topology induced by a family $\{E_\alpha\}$ of Mackey spaces and an accompanying family $\{f_\alpha\}$ of linear operators. To show that E is Mackey it will suffice to prove that the polar $A°$ of each convex, circled, $\sigma(E',E)$-compact subset A of E' is a neighborhood of zero in E . But $A°$ is such a neighborhood iff $(\forall\ \alpha)$ $f_\alpha^{-1}(A°)$ is a neighborhood of zero in E_α . Since $(\forall\ \alpha)$ the transpose f_α^t of f_α is weakly continuous ([R], p.39), $f_\alpha^t(A)$ is $\sigma(E_\alpha', E_\alpha)$-compact as well as circled and convex; consequently, as $f_\alpha^{-1}(A°)$ is just $f_\alpha^t(A)°$ (see

	1. Mackey	2. Infra-barreled	3. Barreled	4. Quasi-complete	5. Semi-reflexive	6. Reflexive	7. Semi-Montel	8. Montel	9. Metrizable	10. Frechet	11. Normed	12. Banach	13. Bornological	14. Ultra-bornological	15. Complete
i. projective topology	-	-	-	-	-	-	-	-	-	-	-	-	-	-	-
ii. product	+	+	+	+	+	+	+	+	-	-	-	-	(+)	(+)	+
iii. projective limit	-	-	-	+	+	-	+	-	-	-	-	-	-	-	+
iv. closed subspace	-	-	-	+	+	-	+	-	+	+	+	+	-	-	+
v. inductive topology	+	+	+	-	-	-	-	-	-	-	-	-	+	+	+
vi. quotient	+	+	+	-	-	-	-	-	+	+	+	+	+	+	-
vii. direct sum	+	+	+	+	+	+	+	+	-	-	-	-	+	+	-
viii. inductive limit	+	+	+	-	-	-	-	-	-	-	-	-	-	+	+
ix. strict inductive limit	+	+	+	+	+	+	+	+	-	-	-	-	+	+	+
x. completion	+	+	+	+					+	+	+	+			+

[R], p.39) and E_α is a Mackey space, we have that $f_\alpha^{-1}(A°) = f_\alpha^t(A)°$ is a neighborhood of zero in E_α.

(vi) See part (v).

(vii) Ibid.

(viii) Ibid.

(ix) Ibid.

(x) [S], p. 133

2. Infrabarreled Spaces.

(i) See (V.F.vii) and (V.F.viii).

(ii) [S], p. 194.

(iii) See (V.A.3iii) and (V.A.5,ii,vii,x).

(iv) See (V.F.viii).

(v) [H], p.218.

(vi) Ibid.

(vii) Ibid.

(viii) Ibid.

(ix) Ibid.

(x) [H], p. 219, Ex.2.

3. Barreled Spaces.

(i) See (V.F.vii) and (V.F.viii).

(ii) [S], p. 138.

(iii) See (V.A.3iii) and (V.A.5,ii,vii,x).

(iv) See (V.F.viii).

(v) [H], p. 214.

(vi) Ibid.

(vii) Ibid.

(viii) Ibid.

(ix) Ibid.

(x) [H], p. 219, Ex. 2.

4. Quasi-complete Spaces.

(i) See (V.F.vii).

(ii) [S], p. 27.

(iii) [S], p. 52.

(iv) Trivial.

(v) See (V.A.4.ix) and (V.A.4.vi).

(vi) Ibid.

(vii) [S], p. 56.

(viii) See (V.A.4.ix) and (V.A.4.vi).

(ix) Each bounded subset of a strict inductive limit is contained in one of the associated subspaces.

(x) Trivial.

5. Semi-reflexive Spaces.

(i) Note that a subspace bears a projective topology and see (V.C.iv).

(ii) [S], p. 146.

(iii) Ibid.

(iv) Immediate.

(v) See (V.A.1.xvi) and (V.A.2.iv)

(vi) Ibid.

(vii) [S], p. 146.

(viii) See (V.H.xi) and (V.H.xii).

(ix) [S], p. 146.

(x)

6. Reflexive Spaces.

(i) See (V.C.iv).

(ii) [S], p. 146.

(iii) See (V.I).

(iv) See (V.A.3.iii) and (V.A.5.x).

(v) See (V.A.1.xvi) and (V.A.2.iv).

(vi) Ibid.

(vii) [S], p. 146.

(viii) See (V.H.xi) and (V.H.xii).

(ix) [S], p. 146.

(x)

7. Semi-Montel Spaces.

(i) (V.C.iv).

(ii) [H], p. 232.

(iii) The projective limit is a closed subspace of a product of semi-Montel spaces.

(iv) [H], p. 232.

(v) See (V.A.1.xvi) and (V.A.2.iv)

(vi) Ibid.

(vii) A bounded subset of a direct sum may be viewed as lying in only a finite direct sum ([S], 11.6.3).

(viii) See (V.S.xi) and (V.H.xii).

(ix) [H], p. 240.

(x) ?

8. Montel Spaces.

(i) See (V.I).

(ii) [S], p. 147.

(iii) See (V.I).

(iv) See (V.A.3.iii) and (V.A.5.ii,vii,x).

(v) See (V.A.1.xvi) and (V.A.2.iv)

(vi) Ibid.

(vii) [S], p. 147.

(viii) See (V.H.xi) and (V.H.xii).

(ix) [S], p. 147.

(x) ?

9. Metrizable Spaces

(i) See (V.C).

(ii) Ibid.

(iii) Ibid. (Since every complete l.c.s. is a projective limit of Banach spaces [S], p. 53).

(iv) Trivial.

(v) See (V.F.vi).

(vi) [S], I.6.3 .

(vii) See (V.F.vi).

(viii) [R], p. 129.

(ix) Ibid.

(x) [H], p. 135.

10. Frechet Space (complete metrizable space).

(i) See (V.C).

(ii) Ibid.

(iii) Ibid. (Since every complete l.c.s. is a projective limit of Banach spaces, [S], p. 53).

(iv) Trivial.

(v) See (V.F.vi)

(vi) [S], I.6.3.

(vii) See (V.F.vi).

(Viii) [R], p. 129.

(ix) Ibid.

(x) Trivial.

11. Normed Spaces.

(i) See (V.C).

(ii) Ibid.

(iii) Ibid. (Since every complete l.c.s. is a projective limit of Banach spaces, [S] p.53).

(iv) Trivial.

(v) See (V.F.vi).

(vi) [S] II.2.3.

(vii) See (V.F.vi).

(viii) [R], p. 129.

(ix) Ibid.

(x) [S], p. 41.

12. Banach Spaces.

(i) See (V.C).

(ii) Ibid.

(iii) Ibid. (Since every complete l.c.s. is a projective limit of Banach spaces. [S] p. 53).

(iv) Trivial.

(v) See (V.F.vi.).

(vi) [S] II.2.3 .

(vii) See (V.F.vi).

(viii) [R], p. 129.

(ix) Ibid.

(x) Trivial.

13. Bornological Spaces.

(i) See (V.A.5.ix,vii,ii) and (V.A.3.xi).

(ii) See [K] 28.8 .

(iii) See (V.A.5.xi,xii) since every complete l.c.s. is a projective limit of Banach spaces. ([S] p. 53).

(iv) (V.A.5.xi,vii,ii) and (V.A.3.xi,xii).

(v) [S], p.61.

(vi) Ibid.

(vii) Ibid.

(viii) Ibid.

(ix) Ibid.

(x)

14. Ultra-bornological Spaces.

(i) See (V.A.5.xi,vii,ii) and (V.A.3.xi).

(ii) The argument given in [K] 28.8 applies here with minor changes.

(iii) See (V.A.5.xi,xii) since every complete l.c.s. is a projective limit of Banach spaces. ([S] p. 53).

(iv) See (V.A.5.xi,vii,ii) and (V.A.3.xi,xii).

(v) [S], II.6.1.

(vi) Ibid.

(vii) Ibid.

(viii) Ibid.

(ix) Ibid.

(x)

15. Complete Spaces.

(i) See (V.F.vii).

(ii) [S], II.5.3 .

(iii) Ibid.

(iv) Elementary.

(v) See (V.A.4.ix) and (V.A.4.vi).

(vi) Ibid.

(vii) [S], II.6.2 .

(viii) See (V.A.4.ix) and (V.A.4.vi).

(ix) [S], II.6.6.

(x) Trivial.

IV. IMPLICATION TABLE

The table below lists various types of locally convex spaces along the top and along the side. If a plus appears in some box, then a space of the type listed in the row of that box is always a space of that type listed on the column of the box. If a minus appears, then not.

1. Mackey Spaces.

(i) Trivial.

(ii) See (V.A.5.xi).

(iii) Since a barreled space is infra-barreled, see (V.A.5.xi).

(iv) See (V.E.vii).

(v) Ibid.

(vi) Since a semi-Montel space is semi-reflexive, see (V.E.vii).

(vii) Ibid.

(viii) Since a barreled space is Mackey, see (V.B.3.xvi) and (V.B.3.xxviii).

(ix) Ibid.

(x) Since a normed space is Mackey, see (V.F.iv).

(xi) Ibid.

(xii) Since a barreled space is Mackey and a metrizable space is bornological, see (V.B.3.xvi) and (V.B.3.xxviii).

(xiii) Ibid.

(xiv) Ibid.

(xv) Ibid.

2. Infra-barreled Spaces.

(i) [H], p. 218.

(ii) Trivial.

(iii) Since a normed space is infra-barreled, see (V.F.iii) and (V.F.v).

(iv) Since a semi-reflexive space is quasi-complete, see (V.F.iii) and (V.F.iv).

15. Frechet	14. Metrizable	13. Banach	12. Normed	11. Complete	10. Quasi-complete	9. Ultra-Bornological	8. Bornological	7. Montel	6. Semi-Montel	5. Reflexive	4. Semi-reflexive	3. Barreled	2. Infra-barreled	1. Mackey	
+	+	+	+	-	-	+	+	+	-	+	-	+	+	+	i. Mackey
+	+	+	+	-	-	+	+	+	-	+	-	+	+	-	ii. Infra-barreled
+	-	+	-	-	-	+	-	+	-	+	-	+	-	-	iii. Barreled
-	-	-	-	-	-	-	-	+	+	+	+	-	-	-	iv. Semi-reflexive
-	-	-	-	-	-	-	-	+	-	+	-	-	-	-	v. Reflexive
-	-	-	-	-	-	-	-	+	+	-	-	-	-	-	vi. Semi-Montel
-	-	-	-	-	-	-	-	+	-	-	-	-	-	-	vii. Montel
+	+	+	+	-	-	+	+	-	-	-	-	-	-	-	viii. Bornological
+	-	+	-	-	-	+	-	-	-	-	-	-	-	-	ix. Ultra-Bornological
+	-	+	-	+	+	-	-	+	+	+	+	-	-	-	x. Quasi-Complete
+	-	+	-	+	-	-	-	-	-	-	-	-	-	-	xi. Complete
-	-	+	+	-	-	-	-	-	-	-	-	-	-	-	xii. Normed
-	-	+	-	-	-	-	-	-	-	-	-	-	-	-	xii. Banach
+	+	+	+	-	-	-	-	-	-	-	-	-	-	-	xiv. Metrizable
+	-	+	-	-	-	-	-	-	-	-	-	-	-	-	xv. Frechet

(v) Ibid.

(vi) Ibid.

(vii) Ibid.

(viii) Since a barreled space is infra-barreled, see (V.B.3.xxviii) and (V.B.3.xvi) .

(ix) Ibid.

(x) Since a normed space is infra-barreled, see (V.F.iv).

(xi) Ibid.

(xii) Since a Montel space is infra-barreled, see(V.C.i) and (V.C.iii) .

(xiii) Ibid.

(xiv) Ibid.

(xv) Ibid.

3. <u>Barreled Spaces</u>.

(i) [S], IV.3.4.

(ii) Trivial.

(iii) Trivial.

(iv) Since Banach spaces are barreled, see (V.E.ii) and (V.E.vii).

(v) Ibid.

(vi) Since Banach spaces are barreled and semi-Montel spaces are semi-reflexive, see (V.E.ii) and (V.E.vii).

(vii) Ibid.

(viii) See (V.B.xxviii) and (V.B.xvi).

(ix) Ibid.

(x) Since ultra-bornological spaces are barreled, see (V.A.4.vi) and (V.A.4.ix).

(xi) Ibid.

(xii) Since Montel spaces are barreled, see (V.C.i) and (V.C.iii).

(xiii) Ibid.

(xiv) Ibid.

(xv) Ibid.

4. <u>Semi-reflexive Spaces</u>

(i) Since a semi-Montel space is semi-reflexive, see (V.D.x) and (V.D.xi).

(ii) Ibid. (Since an infra-barreled space is Mackey).

(iii) Ibid. (Since a barreled space is Mackey).

(iv) Trivial.

(v) Since a reflexive space is infra-barreled, the example of (V.A.5.x) is not reflexive, but a Montel space is semi-reflexive and a closed subspace of a semi-reflexive space is semi-reflexive, so (V.A.5.ii,vii) and (V.A.3.iii) imply that this example is semi-reflexive.

(vi) See (V.D.iii) and (V.D.viii).

(vii) Ibid.

(viii) See (V.B.xxvi) and (V.B.xxviii).

(ix) Ibid.

(x) [S], P. 144.

(xi) See (V.B.xxvi) and (V.B.xxvii).

(xii) Since a Montel space is semi-reflexive, see (V.C.i) and (V.C.iii).

(xiii) Ibid.

(xiv) Ibid.

(xv) Ibid.

5. <u>Reflexive Spaces</u>.

(i) Since a reflexive space is barreled, see [S], IV.3.4.

(ii) Trivial.

(iii) Trivial.

(iv) Trivial.

(v) Trivial.

(vi) See (V.D.iii) and (V.D.viii).

(vii) Ibid.

(viii) See (V.B.xxvi) and (V.B.xxviii).

(ix) Ibid.

(x) [S], p. 144.

(xi) See (V.B.xxvi) and (V.B.xxvii).

(xii) Since a Montel space is reflexive, see (V.C.i) and (V.C.iii).

(xiii) Ibid.

(xiv) Ibid.

(xv) Ibid.

6. Semi-Montel Spaces.

(i) See (V.D.x) and (V.D.xi).

(ii) Ibid. (Since an infra-barreled space is Mackey).

(iii) Ibid. (Since a barreled space is Mackey).

(iv) Trivial.

(v) Since a reflexive space is infra-barreled, the example (V.A.5.x) is not reflexive, but (V.A.5.ii,vii) and (V.A.3.iii) show that it is the closed subspace of a semi-Montel space, and thus semi-Montel.

(vi) Trivial.

(vii) Since a Montel space is infra-barreled, the example of (V.A.5.x) is not Montel; but (V.A.5.ii,vii) and (V.A.3.iii) show that it is the closed subspace of a semi-Montel space, and thus semi-Montel.

(viii) See (V.B.xxvi) and (V.B.xxviii).

(ix) Ibid.

(x) Since a semi-Montel space is semi-reflexive, see [S], p. 144.

(xi) See (V.B.xxvi) and (V.B.xxvii).

(xii) See (V.C.i) and (V.C.iii).

(xiii) Ibid.

(xiv) Ibid.

(xv) Ibid.

7. Montel Spaces.

(i) Since a Montel space is barreled, see [S], IV.3.4.

(ii) Trivial.

(iii) Trivial.

(iv) Trivial.

(v) Trivial.

(vi) Trivial.

(vii) Trivial.

(viii) See (V.B.xxvi) and (V.B.xxviii).

(ix) Ibid.

(x) Since a Montel space is semi-reflexive, see [S], p. 144.

(xi) See (V.B.xxvi) and (V.B.xxvii).

(xii) See (V.C.i) and (V.C.iii).

(xiii) Ibid.

(xiv) Ibid.

(xv) Ibid.

8. Bornological Spaces.

(i) [S], IV.3.4.

(ii) [S], p. 142.

(iii) Since a normed space is bornological, see (V.F.iii) and (V.F.v).

(iv) Since a Banach space is bornological, see (V.E.ii) and (V.E.vii).

(v) Ibid.

(vi) Since a Banach space is bornological and a semi-Montel space is semi-reflexive, see (V.E.ii) and (V.E.vii).

(vii) Ibid.

(viii) Trivial.

(ix) Since a normed space is bornological and an ultra-bornological space is barreled, see (V.F.iii) and (V.F.v).

(x) Since a normed space is bornological, see (V.F.iii) and (V.F.iv).

(xi) Ibid.

(xii) See (V.C.ii) and (V.C.iii).

(xiii) Ibid.

(xiv) Ibid.

(xv) Ibid.

9. Ultra-Bornological Spaces.

(i) [S], IV.3.4.

(ii) [S], p. 142.

(iii) [H], p. 287.

(iv) Since a Banach space is ultra-bornological, see (V.E.ii) and (V.E.vii).

(v) Ibid.

(vi) Since a semi-Montel space is semi-reflexive, see (V.E.ii) and (V.E.vii).

(vii) Ibid.

(viii) Trivial.

(ix) Trivial.

(x) See (V.A.4.vi) and (V.A.4.ix).

(xi) Ibid.

(xii) See (V.C.ii) and (V.C.iii).

(xiii) Ibid.

(xiv) Ibid.

10. Qausi-complete Spaces.

(i) See (V.D.xiii) and (V.D.x).

(ii) Ibid. (Since an infra-barreled space is Mackey).

(iii) Ibid. (Since a barreled space is Mackey).

(iv) See (V.E.ii) and (V.E.vii).

(v) Ibid.

(vi) Ibid. (Since a semi-Montel space is semi-reflexive).

(vii) Ibid. (Since a Montel space is semi-reflexive).

(viii) Since a Montel space is quasi-complete, see (V.B.3.xxvi) and (V.B.3.xxviii).

(ix) Ibid.

(x) Trivial.

(xi) Since a Montel space is quasi-complete, see (V.B.3.xxvi) and (V.B.3.xxvii).

(xii) See (V.C.i) and (V.C.iii).

(xiii) Ibid.

(xiv) Ibid.

(xv) Ibid.

11. Complete Spaces.

(i) See (V.D.xiii) and (V.D.x).

(ii) Ibid. (Since an infra-barreled space is Mackey).

(iii) Ibid. (Since a barreled space is Mackey).

(iv) See (V.E.ii) and (V.E.vii).

(v) Ibid.

(vi) Ibid. (Since a semi-Montel space is semi-reflexive).

(vii) Ibid. (Since a Montel space is semi-reflexive).

(viii) See (V.A.5.xi) and (V.A.5.xii).

(ix) Ibid.

(x) Trivial.

(xi) Trivial.

(xii) See (V.C.i) and (V.C.iii).

(xiii) Ibid.

(xiv) Ibid.

(xv) Ibid.

12. Normed Spaces.

(i) Since a bornological space is Mackey, see [S], II.8.1.

(ii) Since a bornological space is infra-barreled, see [S], II.8.1.

(iii) See (V.F.iii) and (V.F.v).

(iv) See (V.E.ii) and (V.E.vii).

(v) Ibid.

(vi) Ibid. (Since a semi-Montel space is semi-reflexive).

(vii) Ibid. (Since a Montel space is semi-reflexive).

(viii) [S],II.8.1.

(ix) Since an ultra-bornological space is barreled, see (V.F.iii) and (V.F.v).

(x) See (V.F.iii) and (V.F.iv).

(xi) Ibid.

(xii) Trivial.

(xiii) See (V.F.iii) and (V.F.iv).

(xiv) Trivial.

(xv) See (V.F.iii) and (V.F.iv).

13. Banach Spaces.

(i) Since a barreled space is Mackey, see [S], II.7.1.

(ii) Ibid.

(iii) Ibid.

(iv) See (V.E.ii) and (V.E.vii).

(v) Ibid.

(vi) Ibid. (Since a semi-Montel space is semi-reflexive).

(vii) Ibid. (Since a Montel space is semi-reflexive).

(viii) [S], II.8.1.

(ix) Trivial.

(x) Trivial.

(xi) Trivial.

(xii) Trivial.

(xiii) Trivial.

(xiv) Trivial.

(xv) Trivial.

14. Metrizable Spaces.

(i) Since a bornological space is Mackey, see [S], II.8.1.

(ii) Since a bornological space is infra-barreled, see [S], LL.8.1.

(iii) See (V.F.iii) and (V.F.v).

(iv) See (V.E.ii) and (V.E.vii).

(v) Ibid.

(vi) Ibid. (Since a semi-Montel space is semi-reflexive).

(vii) Ibid. (Since a Montel space is semi-reflexive).

(viii) [S], LL.8.1.

(ix) Since an ultra-bornological space is barreled, see (V.F.iii).

(x) See (V.F.iii) and (V.F.iv).

(xi) Ibid.

(xii) See (V.G.iii) and (V.G.iv).

(xiii) See (V.F.iii) and (V.F.iv).

(xiv) Trivial.

(xv) See (V.F.iii) and (V.F.iv).

15. Frechet Spaces.

(i) Since a barreled space is Mackey, see [S], II.7.1.

(ii) Ibid.

(iii) Ibid.

(iv) See (V.E.ii) and (V.E.vii).

(v) Ibid.

(vi) Ibid. (Since a semi-Montel space is semi-reflexive).

(vii) Ibid. (Since a Montel space is semi-reflexive).

(viii) [S], II.8.1.

(ix) [H], P. 287.

(x) Trivial.

(xi) Trivial.

(xii) See (V.G.iii) and (V.G.iv).

(xiii) Ibid.

(xiv) Trivial.

(xv) Trivial.

V. EXAMPLES

In this chapter are collected those examples necessary to substantiate the claims of chapters III and IV.

A. The Eminent Example of Gottfried Köthe .[S, p. 195]

1. A Frechet Montel Space. For each $n \in N$, let $a^{(n)} | N \times N \to K$ be defined by

$$\text{(i)} \qquad a_{ij}^{(n)} \equiv \begin{cases} i^n : \text{for} & i<n \\ n^i : \text{for} & i \geq n \end{cases} \qquad (\forall\ i, j \in N).$$

Then $(\forall\ n \in N)(\forall\ i, j \in N)$

$$\text{(ii)} \qquad a_{ij}^{(n+1)} \geq a_{ij}^{(n)} \text{ and } a_{ij}^{(n+1)} \geq \left(\frac{n+1}{n}\right)^i \cdot a_{ij}^{(n)} \quad \text{if } i>n .$$

Let $(\forall\ n \in N)$

$$\text{(iii)} \qquad E^{(n)} \equiv \{x \in K^{N \times N} : \sum_{i,j \in N} |x_{ij}| \cdot a_{ij}^{(n)} < \infty\}$$

and define $\| \ \|^{(n)} | E^{(n)} \to R$ by

$$\text{(iv)} \qquad \|x\|^{(n)} \equiv \sum_{i,\ j \in N} |x_{ij}| \cdot a_{ij}^{(n)} \qquad (\forall\ x \in E^{(n)}).$$

For each $n \in N$, let μ_n be the measure defined on the power set of $N \times N$ given by

$$\text{(v)} \qquad \mu_n(A) \equiv \sum_{(i,\ j) \in A} a_{ij}^{(n)} \qquad (\forall\ A \subset N \times N).$$

For each $n \in N$,

$$\text{(vi)} \qquad E^{(n)} \text{ is the space } L_1(N \times N, \mu_n) \text{ and } \| \ \|^{(n)} \text{ is the } L_1\text{-norm.}$$

In particular, μ_1 is the counting measure

(vii) $E^{(1)} = \ell_1(N\times N)$ and $\| \|^{(1)}$ is the ℓ_1-norm .

The following easily-verified fact will be used in the sequel:

(viii) $(\forall\ A\subset E^{(1)}: A$ is $\| \|^{(1)}$-compact)

$$\lim_n \sup \{|x_{ij}|:i,j>n;x\in A\} = 0 .$$

From (ii) it follows that $(\forall\ m,\ n\in N : m<n)$

(ix) $E^{(n)} \subset D^{(m)}$ and $\|x\|^m \leq \|x\|^n \quad (\forall\ x\in E^{(n)})$.

Thus if $(\forall\ m,n\in N : m<n)$ $\iota_{mn}\ E^{(n)} \to E^{(m)}$ is defined by

(x) $\iota_{mn}(x) = x\ (\forall\ x\in E^{(n)})$, then ι_{mn} is continuous.

Now let

(xi) $E = \bigcap_{n\in E} E^{(n)}$

and

(xii) $\mathfrak{T}$ be the topology induced by the restrictions of the norms $\| \|^{(n)}$ $(n\in N)$ to E .

If $\pi|E \to \prod_{n\in N} E^{(n)}$ is defined by $(\forall\ x\in E)$

(xiii) $\pi_x(n) = x \quad (\forall\ n\in N)$,

then

(xiv) π is a topological isomorphism of E onto the projective limit $\lim\limits_{\leftarrow} \iota_{mn}\ (E^{(n)})$.

Since the projective limit of Banach spaces is complete [S II.5.3], and since E is metrizable $(d(x,y) \equiv \sum_{n\in N} \|x-y\|^{(n)}/(2^n + 2^n \|x-y\|^{(n)})$ for all $x,y\in E$ gives one metric), it follows that

(xv) E is a Frechet space.

We will show that

(xvi) E is a Montel space.

Let A be any closed bounded subset of E and $\{x^{(n)}\}$ any sequence in A. Evidently $\{x^{(n)}\}$ is bounded in $E^{(1)} = \ell_1(N\times N)$ and so is pointwise bounded on $N\times N$. There is a subsequence $\{x^{\sigma(n)}\}$ of $\{x^{(n)}\}$ with a limit x in the product space $K^{N\times N}$. For each $m\in N$ we have $(\forall\ p,q\in N)$

$$\sum_{i=1}^{p} \sum_{j=1}^{q} |x_{ij}|\ a_{ij}^{(m)} \le \sup_{n=1}^{\infty} \left\{ \sum_{i=1}^{p} \sum_{j=1}^{q} |x_{ij}^{(n)}|\ a_{ij}^{(m)} \right\} \le \sup_{n=1}^{\infty} \| x^{(n)} \|^{(m)} < \infty$$

so that $x\in E^{(m)}$. Fix $m\in N$ and find $\alpha > 0$ such $\|x^{(n)}\|^{(m+1)} < \alpha$ $(\forall\ n\in N)$. Let ε be an arbitrary positive number and choose $p\in N$ such that $p > m$ and $\left(\frac{m}{m+1}\right)^p > \varepsilon/\alpha$. By (ii), we have $(\forall\ n\in N)$

$$\sum_{i=p}^{\infty} \sum_{j=1}^{\infty} |x_{ij}^{\sigma(n)}|\ a_{ij}^{(m)} < \sum_{i=p}^{\infty} \sum_{j=1}^{\infty} |x_{ij}^{\sigma(n)} \cdot a_{ij}^{(m+1)} \left(\frac{m}{m+1}\right)^i \le$$

$$\left(\frac{m}{m+1}\right)^p \|x^{\sigma(n)}\|^{(m+1)} < \varepsilon$$

It follows that $\lim_n \|x^{\sigma(n)}-x\|^{(m)} = 0$, so that $\lim_n x^{\sigma(n)} = x$ in E.

Thus, A is $\mathcal{T}$-compact and E is Montel.

2. The Quotient Space $\ell_1(N)$. Let $\mathfrak{s}|E \to \ell_1(N)$ be defined by $(\forall\ x \in E)$

(i) $$\mathfrak{s}_x(n) \equiv \sum_{i \in N} x_{in} \qquad (\forall\ n \in N).$$

Writing $\|\ \|_1$ for the ℓ_1-norm on $\ell_1(N)$ and recalling (1.vii), we have $\|x\|^{(1)} \geq \|\mathfrak{s}_x\|_1 \quad (\forall\ x \in E)$ and so

(ii) $\mathfrak{s}$ is continuous.

Let y be a arbitrary element of $\ell_1(N)$ and write, as usual, δ for Kronecker's delta function on $N \times N$. Define $x \in K^{N \times N}$ be letting $x_{ij} \equiv y_j\ \delta_{1i}\ (\forall\ i,j \in N)$. Then $x \in \bigcap_{n \in N} E^{(n)} = E$ and $\mathfrak{s}_x = y$. This shows that $\mathfrak{s}$ is onto $\ell_1(N)$ and so, by (2.ii), (1.xv), and the open mapping theorem

(iii) $\mathfrak{s}$ is a topological homomorphism of E onto $\ell_1(N)$.

It follows that

(iv) $\ell_1(N)$ (under the $\|\ \|_1$-topology) is topologically isomorphic to the locally convex quotient space $E/\mathfrak{s}^{-1}(0)$.

3. The Dual Space of E. Every metrizable locally convex space is bornological [S II.8.1] and the strong dual of each bornological space is complete [S VI.6.1]. It follows from (1.xv) that

(i) E' is $\beta(E', E)$-complete

Since E is Montel, its bounded subsets are precisely those subsets of which the closed absolutely convex hulls are compact. This implies

(ii) $\beta(E', E) = \tau(E', E) = \varkappa(E', E)$.

Since the strong dual of a Montel space is Montel [S IV.5.9],

(iii) E' is (E', E)-Montel.

For each $n \in N$ and > 0, let

(iv) $B_r^{(n)} \equiv \{x \in E: \|x\|^{(n)} \leq r\}$.

Then the family

(v) $\{B_r^{(n)}: n \in N, r > 0\}$ constitutes a base for the 0-neighborhood system of E .

The polars of the 0-neighborhoods of E are just the closed, absolutely convex, equicontinuous subsets of E' . [S IV.1.5]. Since E is Montel, it is barreled, and so each $\sigma(E', E)$-bounded subset of E' is equicontinuous [S IV.1.6]. Because the $\sigma(E', E)$-bounded subsets of E' are precisely the $\tau(E', E)$-bounded subsets [S IV.3.2], it follows from (2.ii) that the $\beta(E', E)$-bounded subsets of E' are the equicontinuous subsets. It now follows from (3.v) that

(vi) $(\forall A \subset E': A \text{ is } \beta(E', E)\text{-bounded}) (\exists n \in N, r > 0) A \subset B_r^{(n)\circ}$

For each $n \in N$, let $\eta^{(n)} | E^{(n)'} \to E'$ be defined by

(vii) $\eta^{(n)}(f) \equiv f|_E \quad (\forall f \in E^{(n)'})$.

From (2.vi) we have $E' = \bigcup_{\substack{n \in N \\ r > 0}} B_r^{(n)\circ}$. Since $(\forall n \in N)$ a function in $E^{(n)*}$ is in $E^{(n)'}$ if it is bounded on $B_r^{(n)}$ $(\exists r > 0)$, it follows that $B_r^{(n)\circ} \subset \eta(E^{(n)'})$ $(\forall r > 0)$. These facts imply

(viii) $E' = \bigcup_{n=1}^{\infty} \eta^{(n)}(E^{(n)'})$.

Let $\mathfrak{T}$ be the inductive topology on E' induced by the maps $\eta^{(n)} (n \in N)$ when $(\forall n \in N)$ $E^{(n)'}$ bears the strong topology $\beta(E^{(n)'}, E^{(n)})$. Evidently $(\forall n \in N)$ $\eta^{(n)}$ is continuous when $E^{(n)'}$ and E' bear their respective weak topologies $\sigma(E^{(n)'}, E^{(n)})$ and $\sigma(E', E)$ and so $\eta^{(n)}$ is continuous when $E^{(n)'}$ and E' bear their respective strong topologies $\beta(E^{(n)'}, E^{(n)})$ and $\beta(E', E)$ [S IV.7.4.]. This implies

(ix) $\beta(E',E) \subset \mathfrak{T}$.

Since a Montel space is barreled, and since a barreled strong dual of a metrizable space is always bornological [S IV.6.6], it follows from (1.xv) and (3.iii) that

(x) E' is $\beta(E',E)$-bornological.

Now let A be an arbitrary $\beta(E',E)$-bounded subset of E'. By (3.vi) there exist $r > 0$ and $n \in N$ for which $A \subset B_r^{(n)\circ}$; thus, A is a subset of the image by the function $\eta^{(n)}$ of some $\beta(E^{(n)'},E^{(n)})$-bounded subset of $E^{(n)'}$. It follows that A is $\mathfrak{T}$-bounded. By (3.x), this shows that $\mathfrak{T} \subset \beta(E',E)$ and so, by (3.ix),

(xi) $\beta(E',E)$ is the locally convex inductive topology on E' induced by the maps $\eta^{(n)}$ $(n \in N)$; hence, E' under $\beta(E',E)$ is ultra-bornological.

4. An Ultra-bornological Dense Subspace of E'.

Let $(\forall\ n \in N)$

(i) $$F^{(n)} \equiv \left\{ x \in K^{N\times N} : \lim_{i,j\to\infty} (x_{ij})(a_{ij}^{(n)})^{-1} = 0 \right\}$$

and

(ii) $$H^{(n)} \equiv \left\{ x \in K^{N\times N} : \sup_{i,j\in N} |x_{ij}|\ (a_{ij}^{(n)})^{-1} < \infty \right\} .$$

Define $(\forall\ n \in N)$ $h^{(n)} \mid H^{(n)} \to E^{(n)*}$ by $(\forall\ x \in H^{(n)})$

(iii) $$h_x^{(n)}(y) \equiv \sum_{i,j\in N} x_{ij}y_{ij} \quad (\forall\ y \in E^{(n)}) .$$

Under the discrete topology, $N\times N$ is a locally compact topological space and $(\forall\ n \in N)$ the measure μ_n defined in (1.v) is a regular Borel measure on E . Note that $(\forall\ n \in N)$ the map on $H^{(n)}$ sending each $x \in H^{(n)}$ to $x\cdot a^{(n)}$ is a bijection of $H^{(n)}$ onto $L_\infty(N\times N,\mu_n)$ and that $F^{(n)}$ is mapped onto $C_0(N\times N)$, the space of continuous functions on $N\times N$ vanishing at ∞ . Since $(\forall\, n\in N)$ $L_\infty(N\times N,\mu_n)$ (with the supremum norm) is topologically isomorphic to $L_1(N\times N,\mu_n)'$ [HS 20.20], and since $C_0(N\times N)$ is a norm-closed $\sigma(L_\infty,L_1)$-dense subspace of $L_\infty(N\times N,\mu_n)$, it follows that

(iv) $h^{(n)}(H^{(n)}) = E^{(n)'}$ and $h^{(n)}(F^{(n)})$ is a $\beta(E^{(n)'},E^{(n)})$-closed and $\sigma(E^{(n)'},E^{(n)})$-dense subspace of $E^{(n)}$.

Recall (3.vii) and let

(v) $$F = \bigcup_{n\in N} \eta^{(n)} \circ h^{(n)}(F^{(n)}).$$

The inductive topology on F induced by the maps $\eta^{(n)}|h^{(n)}(F^{(n)})$ $(n\in N)$ is just the restriction to F of the inductive topology $\mathfrak{F}$ on E' induced by the maps $\eta^{(n)}$ $(n\in N)$ ([K] 31.6.1). By (3.xi), this implies that

(vi) The subspace F of E' under the topology $\beta(E',E)$ is ultra-bornological.

As remarked before, $(\forall\, n\in N)$ $\eta^{(n)}$ is weakly continuous; by (4.iv) it follows that $\eta^{(n)} \circ h^{(n)}(F^{(n)})$ is $\sigma(E',E)$-dense in $\eta^{(n)} \circ h^{(n)}(H^{(n)}) = \eta^{(n)}(E^{(n)'})$. By (3.viii) this implies that F is $\sigma(E',E)$-dense in E'. But weakly dense subspaces are $\tau(E',E)$-dense [S IV.3.1] and so, by (3.ii),

(vii) F is $\beta(E',E)$-dense in E' .

Let $x\in K^{N\times N}$ be defined by $x_{ij} \equiv \delta_{1i}$ (δ = Kronecker's delta function). Evidently (see (1.i)) $x\in \bigcap_{n=1}^{\infty} H^{(n)}$ but $x \notin \bigcup_{n=1}^{\infty} F^{(n)}$. Thus $\eta^{(1)} \circ h^{(1)}(x)$ is in E' but not in F and

(viii) $F \neq E'$.

For each $n\in N$, let $x^{(n)} \equiv \delta_{1i} \cdot (\sum_{k=1} \delta_{jk})$. It is evident from (1.viii) that the sequence $\{h^{(1)}(x^{(n)})\}$ converges to $h^{(1)}(x)$ uniformly on compact subsets of $E^{(1)}$, and so $\{\eta^{(1)} \circ h^{(1)}(x^{(n)})\}$ converges to $\eta^{(1)} \circ h^{(1)}(x)$ uniformly on compact subsets of E . By (3.ii) , this implies that $\{\eta^{(1)} \circ h^{(1)}(x^{(n)})\}$ converges to $\eta^{(1)} \circ h^{(1)}(x)$ in the topology $\beta(E', E)$. Since convergent sequences are bounded [S.I.5.1], and since $\eta^{(1)} \circ h^{(1)}(x^{(n)})$ is in F for all $n \in N$, this implies

(ix) F is not quasi-complete (bearing the relativized $\beta(E',E)$-topology).

5. A subspace of E' Isomorphic to $\ell_\infty(N)$.

Recall (2.i) and let

(i) $G \equiv \{f \in E' : f(x) = 0 (\forall\, x \in s^{-1}(0))\}$.

Since G is the polar of $s^{-1}(0)$, it is $\tau(E',E)$-closed. [S IV.1.5 and IV.3.1] and so, by (3.ii),

(ii) G is a $\beta(E',E)$-closed subspace of E' .

The map $\dagger | G \to \ell_1(N)^*$ given by $(\forall\, f \in G)(\forall\, x \in \ell_1(N))$

(iii) $\dagger_f(x) \equiv f(y)\ (\forall\, y \in E : s(y) = x)$

is well-defined and [S IV.4.1] implies

(iv) $\dagger$ is an algebraic isomorphism of G onto $\ell_1(N)'$ and

(v) $\dagger^{-1}$ is continuous when $\ell_1(N)'$ bears the Mackey topology $\tau(\ell_1, \ell_\infty)$ and G the relativized topology $\tau(E',E) = \beta(E',E)$.

we shall show

(vi) $\dagger$ is continuous with respect to the topologies mentioned in (v).

Let $\{f_\alpha\}$ be any net in G which $\beta(E',E)$-converges to 0 and let D be any convex, circled $\sigma(\ell_1,\ell_\infty)$-compact subset of $\ell_1(N)$. Then, by Eberlein's Theorem, [S IV.11.1], D is $\sigma(\ell_1,\ell_\infty)$-sequentially compact. Schur's Theorem states that, in $\ell_1(N)$, sequential convergence is equivalent to weak sequential convergence [Y p.122]. It follows that D is $\| \|_1$-compact. The theorem of Banach-Dieudonné [S IV.6.3] implies the existence of a null sequence $\{x_n\}$ in $\ell_1(N)$ of which D is the absolutely convex hull. Since E is metrizable, there is a quasi-norm (or pseudo-norm) $\| \|$ on E which generates the topology of E [S I.6.1]. Let $\| \|_q$ be the function on $\ell_1(N)$ defined by $\|x\|_q = \inf\{\|y\|: y\in \mathfrak{s}^{-1}(x)\}$ $(\forall\, x\in\ell_1(N))$. Then $\| \|_q$ generates the quotient topology on $\ell_1(N)$ [S I.6.3] which, by (2.iv), is just the $\| \|_1$-topology. It follows that there is a null sequence $\{y_n\}$ in E for which $\mathfrak{s}(y_n) = x_n$ $(\forall\, n\in N)$. Since sequences are bounded, the closed, absolutely convex hull W of the set $\{y_n: n\in N\}$ is bounded. Then $\{f_\alpha\}$ converges to 0 uniformly on W and so $\{\dagger_{f_\alpha}\}$ converges to 0 uniformly on $\mathfrak{s}(W)$. But D is a subset of $\mathfrak{s}(W)$ so $\{\dagger_{f_\alpha}\}$ converges to 0 uniformly on D . This proves (5.vi).

Taking (5.vi) together with (5.v), we have

(vii) G, under the relativized topology $\beta(E',E)$, is topologically isomorphic to $\ell_\infty(N)$ under the Mackey topology $\tau(\ell_\infty,\ell_1)$.

As it was for (1.viii), it is easy to verify that

(viii) $(\forall\, A\subset \ell_1(A)$: A is $\| \|_1$-compact) $\lim_{n\to\infty} \sup\{|x(n)|: x\in A\} = 0$.

In the course of proving (5.vi) we observed that

(ix) $\sigma(\ell_1,\ell_\infty)$-compact subsets of $\ell_1(N)$ are $\| \|_1$-compact.

Let $(\forall\, n\in N)$ $x^{(n)} \in \ell_\infty(N)$ be defined by $x^{(n)}(j) \equiv \delta_{nj}$ $(\forall\, j\in N)$. In view of (5.viii) and (5.ix), the sequence $\{x^{(n)}\}$ converges to 0 uniformly on $\sigma(\ell_1,\ell_\infty)$-compact subsets of $\ell_1(N)$; thus $\lim_n x^{(n)} = 0$ in $\tau(\ell_\infty,\ell_1)$. But, since $\beta(\ell_1,\ell_\infty)$ is the norm topology on $\ell_1(N)$, $\{x^{(n)}\}$ does not converge to 0 uniformly on $\beta(\ell_1,\ell_\infty)$-bounded subsets of $\ell_1(N)$. Hence

(x) $\ell_\infty(N)$ is not $\tau(\ell_\infty,\ell_1)$-infrabarreled.

Since a space is infrabarreled if it is either barreled or bornological [S IV.5.2], we have

(xi) $\ell_\infty(N)$ under the topology $\tau(\ell_\infty,\ell_1)$ is neither barreled nor bornological.

By (5.ii), (3.i), and (5.vii)

(xii) $\ell_\infty(N)$ under the topology $\tau(\ell_\infty,\ell_1)$ is complete.

B. The Eminent Example of Yokio Komura. [YK]

1. Notation. Let $A(0) \equiv R$ and $(\forall\ n \in N)$

(i) $A(n) \equiv \{B: B \subset A(n-1)\}$.

The aspect of $A(n)$ which is of interest here is its cardinality: $\overline{\overline{A(n)}} = 2^{\overline{\overline{A(n-1)}}}$. Elements of $\bigcup_{n \in Z^+} A(n)$ will be represented by lower case Roman letters: x, y, w, ...

Let $(\forall\ n \in Z^+)$

(ii) $F^{(n)} \equiv \{\alpha \in R^{A(n)}: \alpha$ has finite support$\}$;

(iii) $C^{(n)} \equiv \{\alpha \in R^{A(n)}: \alpha$ has countable support$\}$.

Elements of $\bigcup_{n \in Z^+} R^{A(n)}$ will be represented by lower case Greek letters: $\alpha, \beta, \gamma, \ldots$.

2. Hamel Bases and Linear Operators. From the appendix on cardinality, we know that the linear dimension of $F^{(0)}$ and $C^{(0)}$ is $\overline{\overline{A(0)}}$; $(\forall\ n \in N)$ the linear dimension of $R^{A(n)}$, of $C^{(n+1)}$, and $F^{(n+1)}$ is $\overline{\overline{A(n+1)}}$. We shall select appropriate Hamel bases.

For each $n \in Z^+$ and $x \in A(n)$, let $\xi_x^{(n)}$ be the characteristic function defined on $A(n)$ of the singleton $\{x\}$. Then $(\forall\ n \in N)$

(i) $B_F^{(n)} \equiv \{\xi_x : x \in A(n)\}$ is a Hamel basis for $F^{(n)}$ for, in fact,

(ii) $$\beta = \sum_{x \in A(n)} \beta(x) \cdot \xi_x \qquad (\forall\ \beta \in F^{(n)}) \ .$$

Let $(\forall\ n \in Z^+)$

(iii) $B_C^{(n)}$ be a Hamel base for $C^{(n)}$ such that $B_F^{(n)} \subset B_C^{(n)}$.

For each $n \in Z^+$, let $<^n$ be a well-ordering on $A(n)$ such that $(\forall\ x \in A(n))$ the set $\{y \in A(n) : y <^n x\}$ has cardinality strictly less than $\overline{\overline{A(n)}}$. For all $x, y \in A(n)$ such that $x <^n y$, let

(iv) $[x,y] \equiv \{x,y\} \cup \{w \in A(n) : \ x <^n w <^n y\}$.

Let $(\forall\ n \in Z^+)$

(v) $L(n) \equiv \{x \in A(n) : \ (\forall\ y \in A(n) : y <^n x)(\exists\ w \in A(n))\ x <^n w <^n y\}$.

Then $\overline{\overline{L(n)}} = \overline{\overline{A(n)}}$ $(\forall\ n \in Z^+)$ and there is a partition $\{L'(n,x) : x \in A(n)\}$ of $L(n)$ such that $(\forall\ x \in A(n))$ $\overline{\overline{L'(n,x)}} = \overline{\overline{A(n)}}$ and $(\forall\ y \in A(n) : y \neq x)$ $L'(n,x) \cap L'(n,y) = \Phi$. Let $(\forall\ n \in Z^+)$ $(\forall\ x \in A(n))$ $L(n,x) \equiv \{y \in L'(n,x) : \ x <^n y\}$; then

(vi) $\overline{\overline{L(n,x)}} = \overline{\overline{A(n)}}$ and

(vii) $L(n,x) \cap L(n,y) = \Phi \qquad (\forall\ y \in A(n) : \ y \neq x)$.

For $n \in Z^+$, $x \in A(n)$, and $y \in L(n,x)$, let $\xi^{(n)}_{[x,y]}$ be the characteristic function defined on $A(n)$ of the set $[x,y]$. Evidently $(\forall\ n \in Z^+)$ the set $\{\xi^{(n)}_{[x,y]} : x \in A(n),\ y \in L(n,x)\}$ is linearly independent. Choose $(\forall\ n \in Z^+)$

(viii) a Hamel base $B^{(n)}$ for $R^{A(n)}$ such that

$(\forall\ x \in A(n))\ (\forall\ y \in L(n,x))\ \xi^{(n)}_{[x,y]} \in B^{(n)}$.

Let $(\forall\ n\in Z^+)$

(ix) $\Psi_{FC}^{(n)}$ be a linear isomorphism of $F^{(n)}$ onto $C^{(n)}$ such that

$$\Psi_{FC}^{(n)}[B_F^{(n)}] = B_C^{(n)}$$

and let

(x) $\Psi^{(n)}$ be a linear isomorphism of $F^{(n+1)}$ onto $R^{A(n)}$ such that $\Psi^{(n)}[B_F^{(n+1)}] = B^{(n)}$.

For $n\in Z^+$, $\alpha\in F^{(n)}$, and $\beta\in R^{A(n)}$, define

(xi) $$\langle\alpha,\beta\rangle_n \equiv \sum_{x\in A(n)} \alpha(x) \cdot \beta(x) \ .$$

Let $(\forall\ n\in Z^+)$ $\Psi^{(n)\sim} \mid F^{(n)} \to R^{A(n+1)}$ be defined by $(\forall\ \alpha\in F^{(n)})$ $(\forall\ x\in A(n+1))$

(xii) $$[\Psi^{(n)\sim}(\alpha)](x) \equiv \langle\alpha,\Psi^{(n)}(\xi_x^{(n+1)})\rangle_n$$

For $n\in Z^+$, $\alpha\in F^{(n)}$, and $\beta\in F^{(n+1)}$, (2.xii) and (2.ii) imply

$$\sum_{x\in A(n+1)} \beta(x)\cdot[\Psi^{(n)\sim}(\alpha)](x) = \sum_{x\in A(n+1)} \beta(x)\cdot \sum_{y\in A(n)} \alpha(y)\cdot[\Psi^{(n)}(\xi_x^{(n+1)})](y) =$$

$$\sum_{y\in A(n)} \alpha(y)\cdot \sum_{x\in A(n+1)} \beta(x)\cdot[\Psi^{(n)}(\xi_x^{(n+1)})](y) = \sum_{y\in A(n)} \alpha(y)\cdot[\Psi^{(n)}(\beta)](y).$$

In other words,

(xiii) $$\langle\beta,\Psi^{(n)\sim}(\alpha)\rangle_{n+1} = \langle\alpha,\Psi^{(n)}(\beta)\rangle_n \ (\forall\ n\in Z^+,\ \alpha\in F^{(n)},\ \beta\in F^{(n+1)}) ,$$

We proceed to show $(\forall\ n\in Z^+)$

(xiv) $\Psi^{(n)\sim}$ is a linear isomorphism of $F^{(n)}$ into $R^{A(n+1)}$ and

(xv) $\Psi^{(n)\sim}[F^{(n)}] \cap C^{(n+1)} = \{0\}$.

Let $n \in Z^+$ and $\alpha \in F^{(n)}$ be non-zero. Choose $x \in A(n)$ such that $\alpha(x) \neq 0$ $\alpha(w) = 0$ $(\forall\ w \in A(n) : x <^n w)$. By (2.vi), the set $L(n,x)$ is uncountable. If $S \equiv \{z \in A(n+1) : (\exists\ y \in L(n,x))\ \Psi^{(n)}(\xi_z^{(n+1)}) = \xi_{[x,y]}^{(n)}\}$, then (2. viii) and (2.x) imply that S is uncountable. For each $z \in S$ and $y \in L(n,x)$ for which $\Psi^{(n)}(\xi_z^{(n+1)}) = \xi_{[x,y]}^{(n)}$, (2.xii) and (2.xi) imply

$$[\Psi^{(n)\sim}(\alpha)](z) = \sum_{w \in A(n)} \alpha(w) \cdot [\Psi^{(n)}(\xi_z^{(n+1)})](w) =$$

$$\sum_{w \in A(n)} \alpha(w) \cdot \xi_{[x,y]}^{(n)}(w) = \sum_{w \in [x,y]} \alpha(w) = \alpha(x) \neq 0 .$$

This shows that $\Psi^{(n)\sim}(\alpha)$ is not in $C^{(n+1)}$. That proves (2.xiv) and (2.xv).

3. The Locally Convex Space F . Let

(i) $$F \equiv \prod_{n \in Z^+} F^{(n)} .$$

Elements of $\prod_{n \in Z^+} R^{A(n)}$ will be represented by lower case German script: $\mathfrak{a}, \mathfrak{b}, \mathfrak{c}, \ldots$.

Define the function $\mathfrak{J}\,|F \to \prod_{n \in Z^+} R^{A(n)}$ by letting $(\forall\ \mathfrak{a} \in F)$

(ii) $$\mathfrak{J}(\mathfrak{a})_0 \equiv \Psi_{FC}^{(0)}(\mathfrak{a}_0) \text{ and}$$

(iii) $$\mathfrak{J}(\mathfrak{a})_n \equiv \Psi_{FC}^{(n)}(\mathfrak{a}_n) + \Psi^{(n-1)\sim}(\mathfrak{a}_{n-1})\ (\forall\ n \in N).$$

We proceed to show that

(iv) $\mathfrak{F}$ is a linear isomorphism of F into $\prod_{n\in Z^+} R^{A(n)}$.

Let $\alpha \in F$ be non-zero and choose $n\in Z^+$ for which $\alpha_n \neq 0$. Then (2.ix) implies that $\Psi_{FC}^{(n)}(\alpha_n) \neq 0$. If $n = 0$, then $\mathfrak{F}(\alpha)_0 = \Psi_{FC}^{(0)}(\alpha) \neq 0$.

If $n \neq 0$, then $\mathfrak{F}(\alpha)_n = \Psi_{FC}^{(n)}(\alpha_n) + \Psi^{(n-1)^\sim}(\alpha_{n-1})$ which, by (2.xv), is non-zero. Hence, $\mathfrak{F}(\alpha) \neq 0$. That (3.iv) holds is now evident.

Let

(v) $\mathfrak{T}$ be the coarsest topology on F for which $\mathfrak{F}$ is continuous $\prod_{n\in Z^+} R^{A(n)}$ bearing the product topology.

Thus, if F bears $\mathfrak{T}$

(vi) $\mathfrak{F}$ is a topological isomorphism.

For each $\alpha \in \oplus_{n\in Z^+} F^{(n)}$, define $f_\alpha \mid \prod_{n\in Z^+} R^{A(n)} \to R$ by

(vii) $f_\alpha(b) \equiv \sum_{n\in Z^+} \langle \alpha_n, b_n \rangle_n \quad (\forall\, b \in \prod_{n\in Z^+} R^{A(n)})$.

As is well-known (see [S IV.4.3]), the dual of the locally convex product $\prod_{n\in Z^+} R^{A(n)}$ is just $\{f_\alpha : \alpha \in \oplus_{n\in Z^+} F^{(n)}\}$. Thus, (3.vi) implies (where F bears $\mathfrak{T}$)

(viii) $F' = \{f_\alpha \circ \mathfrak{F} : \alpha \in \oplus_{n\in Z^+} F^{(n)}\}$.

Since $\prod_{n\in Z^+} R^{A(n)}$ bears its weak topology, it follows from (3.vi) that

(ix) $\mathfrak{T} = \sigma(F, F')$.

Our next goal is (3.xi) below. To this end, we first show

(x) if D is a $\sigma(F' \ F)$-bounded subset of F', then

$$(\exists \ m\epsilon Z^+) \ (\forall \ j\epsilon N: j > m) \ (\forall \ f_\alpha \circ F\epsilon D: \alpha \epsilon \underset{n\epsilon Z^+}{\oplus} F^{(n)}) \ a_j = 0 \ .$$

Let D be as in (3.x) . Assume that (3.x) is false. Chose an increasing sequence $\{s(n)\}$ in N and a sequence $\{a^{(n)}\}$ in $\underset{n\epsilon Z^+}{\oplus} F^{(n)}$ such that $(\forall \ n\epsilon N) \ f_\alpha(n) \circ \Im \epsilon D$, $a_{s(n)}^{(n)} \neq 0$, and $(\forall \ j\epsilon N: j > s(n)) \ a_j^{(n)} = 0$.

Define $b\epsilon F$ inductively as follows: let $b_0 = 0$ and suppose that b_{n-1} is defined; if $n \not\epsilon s(N)$, let $b_n \equiv 0$ and, if $(\exists \ j\epsilon N) \ s(j) = n$, apply (2.ix) to obtain $b_n\epsilon F^{(n)}$ such that

$$| \langle a_{s(j)}^{(j)} \ , \ \Psi_{FC}^{(s(j))}(b_{s(j)}) \rangle_{s(j)} | \geq$$

$$n + | \langle a_{s(j)}^{(j)}, \ \Psi^{(s(j)-1)\sim}(b_{s(j)-1}) \rangle_{s(j)} +$$

$$\sum_{k=1}^{s(j)-1} \langle a_k^{(j)} \ , \ \Psi_{FC}^{(k)}(b_k) + \Psi^{(k-1)\sim}(b_{k-1}) \rangle_k | \ .$$

This, with (3.iii) yields $(\forall \ j\epsilon N)$

$$|f_\alpha(n) \circ \Im(b)| = |\sum_{k=0}^{\infty} \langle a_k^{(j)} \ , \ \Im(b)_k \rangle_k| =$$

$$|\sum_{k=1}^{s(j)} \langle a_k^{(j)} \ , \ \Psi_{FC}^{(k)}(b_k) + \Psi^{(k-1)\sim}(b_{k-1}) \rangle_k| \geq$$

$$|\langle a_{s(j)}^{(j)} \ , \ \Psi_{FC}^{(s(j))}(b_{s(j)}) \rangle_{s(j)}|-$$

$$|\langle a_{s(j)}^{(j)} \ , \ \Psi^{(s(j)-1)\sim}(b_{s(j)-1}) \rangle_{s(j)} +$$

$$\sum_{k=1}^{s(j)-1} \langle \mathfrak{a}^{(j)}_{k} , \Psi^{(k)}_{FC}(\mathfrak{b}_k) + \Psi^{(k+1)\sim}(\mathfrak{b}_{k-1}) \rangle| \geq$$

$$n .$$

But this implies that $\{ f_{\mathfrak{a}(n)} \circ F \}$ and thus D are unbounded: a contradiction. This proves (3.x).

We now show

(xi) each $\sigma(F',F)$-bounded subset D of F' is finite dimensional.

Let D be as in (3.xi). Assume that (3.xi) is false. In view of (3.x), $(\exists\, m\in Z^+)$ such that $\{\mathfrak{a}_m : \mathfrak{a} \in \oplus_{n\in Z^+} F^{(n)} , f_{\mathfrak{a}} \circ \mathfrak{F} \in D\}$ is an infinite dimensional subset of $F^{(m)}$. Since $F^{(m)}$ may be identified with the dual of the locally convex product space $R^{A(m)}$ (see [S] IV.4.3), and since the only bounded subsets of this dual are finite dimensional, it follows that $(\exists\, \alpha\in R^{A(m)})$

(xii) $\sup\{|\langle\mathfrak{a}_m,\alpha\rangle_m| : \mathfrak{a} \in \oplus_{n\in Z^+} F^{(n)} , f_{\mathfrak{a}} \circ\mathfrak{F} \in D\} = \infty$

We shall now change the definition of m slightly. Because of (3.x), we may define m to be the largest integer in N such that there exists $\alpha\in R^{A(m)}$ for which (3.xii) holds. Choose a sequence $\{\mathfrak{a}^{(n)}\}$ such that $(\forall\, n\in N)$ $f_{\mathfrak{a}(n)} \circ \mathfrak{F} \in D$ and

(xiii) $\sup\{|\langle\mathfrak{a}^{(n)}_m,\alpha\rangle_m| : n\in N\} = \infty$.

Since $\{\mathfrak{a}^{(n)}_m\}$ is a sequence in $F^{(m)}$, it is evident that there exists $\beta\in C^{(m)}$ such that $(\forall\, n\in N)$ $\langle\mathfrak{a}^{(n)}_m ,\beta\rangle_m = \langle\mathfrak{a}^{(n)}_m,\alpha\rangle_m$. Since $\Psi^{(m)\sim}\circ\Psi^{(m)^{-1}}_{FC}(\beta)$ is in $R^{A(m+1)}$, it follows by the choice of m that

(xiv) $M \equiv \sup\{| \langle \mathfrak{a}^{(n)}_{m+1} , \Psi^{(m)\sim}\circ\Psi^{(m)^{-1}}_{FC}(\beta)\rangle_{m+1}|\} < \infty$.

By (xiii), there exists a subsequence $\{a^{s(n)}\}$ of $\{a^{(n)}\}$ such that
($\forall\ n\epsilon N$)

(xv) $\qquad |\langle a_m^{s(n)},\beta\rangle| = |\langle a_m^{s(n)},\alpha\rangle| > M+n$.

Define $b\ \epsilon F$ by letting $b_j \equiv 0$ $(\forall\ j\epsilon Z^+:\ j \neq m)$ and $b_m \equiv \Psi_{FC}^{(m)}{}^{-1}(\beta)$.
Then $(\forall\ n\epsilon N)$ (3.vii), (3.ii), (3.iii), (3.xiv), and (3.xv) imply

$$|f_{a^{s(n)}} \circ \mathfrak{I}(b)| = |\sum_{k=0}^{\infty} \langle a_k^{s(n)}, \mathfrak{I}(b)_k\rangle_k| =$$

$$|\langle a_0^{s(n)}, \Psi_{FC}^{(0)}(b_0)\rangle_0 + \sum_{k=1}^{\infty} \langle a_k^{s(n)}, \Psi_{FC}^{(k)}(b_k) + \Psi^{(k-1)\widetilde{}}(b_{k-1})\rangle_k| =$$

$$|\langle a_m^{s(n)}, \Psi_{FC}^{(m)}(b_m)\rangle_m + \langle a_{m+1}^{s(n)}, \Psi^{(m)\widetilde{}}(b_m)\rangle_{m+1}| =$$

$$|\langle a_m^{s(n)},\beta\rangle_m + \langle a_{m+1}^{s(n)}, \Psi^{(m)\widetilde{}} \circ \Psi_{FC}^{(m)}(\beta)\rangle_{m+1}| \geq$$

$$|\langle a_m^{s(n)},\beta\rangle_m| - M \geq n+M-M = n \qquad .$$

This implies that $\{f_{a^{s(n)}} \circ \mathfrak{I}\}$ and thus D are unbounded: a contradiction. This proves (3.xi).

Since every finite dimensional bounded subset of a dual is equicontinuous, it follows from (3.xi) that the equicontinuous subsets of F' are precisely the $\sigma(F',F)$-bounded subsets. That is,

(xvi) $\qquad$ F, under $\mathfrak{T}$, is barreled.

We shall now show

(xvii) $\qquad$ for any $\sigma(F,F')$-bounded subset H of F and $(\forall\ n\epsilon Z^+)$
$\qquad\qquad \{\mathfrak{I}(a)_n:\ a\epsilon H\}$ is finite dimensional.

For any subset H of $\prod_{n\in Z^+} R^{A(n)}$, let

(xviii) $$H_n \equiv \{\alpha_n : \alpha \in H\} .$$

In view of (2.xv), (2.ix), and (3.iii), we have the following diagram for any $H \subset F$ (where " $\leftrightarrow$ " means "bijection"):

$$\mathfrak{F}(H)_0 = \Psi_{FC}^{(0)}(H_0)$$

(xix) $$\mathfrak{F}(H)_n \subset \Psi_{FC}^{(n)}(H_n) \oplus \Psi^{(n-1)\sim}(H_{n-1}) \qquad (\forall\ n\in N);$$

$$\Psi^{(n)\sim}(H_n) \leftrightarrow H_n \leftrightarrow \Psi_{FC}^{(n)}(H_n) \qquad (\forall\ n\in Z^+) .$$

Assume that (3.xvii) is false. Then there exists a $\sigma(F,F')$-bounded subset H of F and an integer $m\in Z^+$ such that H_m is infinite dimensional. By (3.xix), it is evident that m may be chosen such that $\Psi^{(m)\sim}(H_m)$ is infinite dimensional. Choose a sequence $\{\alpha^{(n)}\}$ from H such that $\{\Psi^{(m)\sim}(\alpha_m^{(n)})\}_{n=1}^{\infty}$ is linearly independent in $R^{A(m+1)}$. Since $(\forall\ n\in N)$ the support of $\Psi_{FC}^{(m+1)}(\alpha_{m+1}^{(n)})$ is countable, there exists a countable subset $x = \{x_j\}_{j=1}^{\infty}$ of $A(m+1)$ such that

(xx) $$[\Psi_{FC}^{(m+1)}(\alpha_{m+1}^{(n)})](x) = 0 \qquad (\forall\ n\in N) \quad (\forall\ x\in A(m+1): x\notin X) .$$

Choose a decreasing sequence $\{s(n)\}$ of positive numbers as follows: let $s(1) \equiv 1$ and, having chosen $s(n-1)$, let

(xxi) $$s(n) < \min_{k=1}^{n-1} \frac{|s(k) \cdot [\Psi_{FC}^{(m+1)}(\alpha_{m+1}^{(k)})](x_k)|+1}{\max_{j=1}^{n}|[\Psi_{FC}^{(m+1)}(\alpha_{m+1}^{(j)})|+1} .$$

Let $(\forall\ k \in N)$

(xxii) $$c_k \equiv \max_{n=1}^{k} |s(n) \cdot [|\,{}^{(m+1)}_{FC}(a^{(n)}_{m+1}](x_k) + 1\ .$$

For each x_k in X

$$\sup_{n=1}^{\infty} |s(n) \cdot [\Psi^{(m+1)}_{FC}(a^{(n)}_{m+1})](x_k)| \leq$$

by (3.xxii)

$$c_k + \sup_{n=k+1}^{\infty} s(n)\ (|[\Psi^{(m+1)}_{FC}(a^{(n)}_{m+1})](x_k)| + 1) \leq$$

by (3.xxi)

$$c_k + |s(k) \cdot [\Psi^{(m+1)}_{FC}(a^{(k)}_{m+1})](x_k)| + 1 \leq 2c_k$$

It follows now from (3.xx) that

(xxiii) $\{s(n) \cdot \Psi^{(m+1)}_{FC}(a^{(n)}_{m+1})\}$ is a bounded subset of $R^{A(m+1)}$.

From (3.ix) and (3.vi), it follows that $\mathfrak{J}(H)$ is bounded in

$\prod_{n \in Z^+} R^{A(n)}$; in particular,

(xxiv) $\mathfrak{J}(H)_{m+1}$ is bounded in $R^{A(m+1)}$.

For each $n \in N$, (3.iii) implies

$$s(n) \cdot \mathfrak{J}(a^{(n)})_{m+1} = s(n) \cdot \Psi^{(m+1)}_{FC}(a^{(n)}_{m+1}) + s(n) \cdot \Psi^{(m)\sim}(a^{(n)}_{m})\ .$$

This fact, the fact that $\{\Psi^{(m)\sim}(a^{(n)}_{m}) : n \in N\}$ was chosen to be linearly independent, (3.xxiii) and (3.xxiv) imply

(xxv) $\{s(n) \cdot \Psi^{(m)\sim}(a_m^{(n)}): n\epsilon N\}$ is a bounded subset of $R^{A(m+1)}$

which is also linearly independent.

This means that $(\forall \ \alpha\epsilon R^{(m+1)})$

$$\sup\{|\ \langle\alpha,\Psi^{(m)\sim}(a_m^{(n)})\rangle_{m+1}|:\ n\epsilon N\} < \infty$$

which, by (2.xiii) and (2.x) , implies $(\forall \ \beta\epsilon R^{A(m)})$

$$\sup\{|\langle a_m^{(n)},\beta\rangle_m|:\ n\epsilon N\} < \infty \ .$$

This means (see [S] IV.4.3 and IV.3.2) that $\{a_m^{(n)}: n\epsilon N\}$ is bounded in the locally convex direct sum topology on $F^{(m)}$. But (3.xxv) implies that $\{a_m^{(n)}: n\epsilon N\}$ is infinite dimensional and, by [S] II.6.3, this is impossible. This proves (3.xvii).

Every bounded, finite dimensional set is relatively compact, and products of relatively compact sets are relatively compact. It follows from (3.xvii) that every $\sigma(F,F')$-bounded subset of F is contained in a relatively compact set. Hence, by (3.ix), F is semi-Montel. By (3.xvi)

(xxvi)   F, under $\Im = \sigma(F,F')$, is a Montel space.

But, as we proceed to show,

(xxvii)   F is not $\Im$-complete: $\Im$ is a topological isomorphism

onto a proper dense subspace of $\displaystyle\prod_{n\epsilon Z^+} R^{A(n)}$.

That $\Im$ is a topological isomorphism is just (3.vi). Let a be any element of $\displaystyle\prod_{N\epsilon Z^+} R^{A(n)}$, ϵ any positive number, and m any integer in Z^+ . For each $n\epsilon Z^+$ such that $n \leq m$, let $W^{(n)}$ be any finite subset of A(n). By (2.ix) there exists $\alpha_0 \epsilon F^{(0)}$ such that

$(\forall \ x\epsilon W^{(0)}) \ [\Psi_{FC}^{(0)}(\alpha_0)](x) = a_0(x)$. Having, for $n \leq m$, chosen

$\alpha_{n-1} \in F^{(n-1)}$, choose $\alpha_n \in F^{(n)}$ such that $(\forall\ x \in W^{(n)})$ $[\Psi_{FC}^{(n)}(\alpha_n)](x) = \alpha_n(x) - [\Psi^{(n-1)\sim}(\alpha_{n-1})](x)$. Let $b \in F$ be such that $b_n = \alpha_n$ for all $n = 0,1\ldots,m$. From (3.ii) and (3.iii) it is evident that $(\forall\ n = 0,1,\ldots,m)$

$$|[\mathfrak{I}(b)]_n(x) - \alpha_n(x)| = 0 < \varepsilon \qquad (\forall\ x \in W^{(n)}) .$$

It follows that $\mathfrak{I}(F)$ is dense in $\prod_{n \in Z^+} R^{A(n)}$. This proves (3.xxvii).

Finally, we shall show

(xxviii) F, under $\mathfrak{I}$, is not bornological.

In view of the fact that $\mathfrak{I}(F)$ is dense in $\prod_{N \in Z^+} R^{A(n)}$, it is evident that there exists a net $\{a^{(\alpha)}\}$ in F such that $\{\mathfrak{I}(a^{(\alpha)})\}$ tends to 0 but that $(\forall\alpha)$ $(\exists\ x \in A(0))$ $|a_0^{(\alpha)}(x)| > 1$. Thus, if we assign $(\forall\ n \in Z^+)$ $R^{A(n)}$ the topology of uniform convergence, the linear operator $\mathfrak{I}$ is not continuous when $\prod_{n \in Z^+} R^{A(n)}$ bears the product topology. Since there is only one locally convex topology on finite dimensional spaces, it is evident from (3.xvii) that $\mathfrak{I}$ preserves bounded sets. This proves (3.xxviii).

C. <u>The K-Valued Functions on the Continuum.</u>

The scalar field K is a Montel Banach space. It follows that

(i) K^R is a complete Montel space.

Since $\overline{\overline{R}}$ is not a strongly inaccessible cardinal, and since K is ultra-bornological,

(ii) K^R is ultra-bornological.

However, since $\overline{\overline{R}} > \aleph_0$, K^R evidently does not have a countable neighborhood base for zero; that is,

(iii) K^R is not metrizable.

Let $B \equiv \{f \in K^R: f(x) = 0$ for all but a finite number of x in $R\}$. Then, if $A \equiv \{f \in K^R: (\forall\, x \in R)\ |f(x)| \leq 1\}$, $B \cap A$ is dense in A and A is weakly compact. Hence,

(iv) B is not semi-reflexive.

D. The Space $\ell_2(N)$.

By $\ell_2(N)$ is meant the linear space

(i) $$\{x \in K^N: \sum_{j=1}^{\infty} |x_j|^2 < \infty\} .$$

The norm $\|\ \|_2$ on $\ell_2(n)$ is defined by

(ii) $$\|x\|_2 \equiv \left(\sum_{j=1}^{\infty} |x_j|^2\right)^{\frac{1}{2}} .$$

Denote the topology generated by $\|\ \|_2$ by $\mathfrak{J}_2$. Then

(iii) $\ell_2(N)$ is a reflexive Banach space under $\mathfrak{J}_2$.

In fact, as is well-known, $\ell_2(N)$ is a Hilbert space.

For each $x \in \ell_2(N)$, define the semi-norm $\|\ \|_x$ by

(iv) $$\|y\|_x \equiv \sum_{n=1}^{\infty} |x_n \cdot y_n| .$$

The topology on $\ell_2(N)$ generated by the family $\{\|\ \|_x : x \in \ell_2(N)\}$ of semi-norms is called the normal topology on $\ell_2(N)$ and will be denoted by $\mathfrak{J}_\eta$. For each x and y in $\ell_2(N)$, Holder's Inequality implies

(v) $$\|y\|_x \leq \|y\|_2\ \|x\|_2 .$$

Let f be any $\| \|_2$-continuous linear functional on $\ell_2(N)$. The representation theorem of F. Riesz implies that $(\exists\, x \in \ell_2(N))(\forall\, y \in \ell_2(N)) f(y) = \sum_{n=1}^{\infty} \bar{x}_n y_n$. It follows from (iv) that

$$|f(y)| = |\sum_{n=1}^{\infty} \bar{x}_n y_n| \le \|y\|_x \qquad (\forall\, y \in \ell_2(N)) .$$

Consequently, f is $\mathfrak{T}_\eta$-continuous. It follows from this and (v) that

(vi) $\mathfrak{T}_\eta$ is a topology on $\ell_2(N)$ finer than the weak topology (determined by $\mathfrak{T}_2$) and coarser than $\mathfrak{T}_2$.

Consequently ([S IV p. 132]) ,

(vii) the $\mathfrak{T}_\eta$ bounded subsets of $\ell_2(N)$ are just the $\mathfrak{T}_2$-bounded subsets.

Write B for the closed $\| \|_2$-unit ball of $\ell_2(N)$. For each $m \in N$, let $w^{(m)}$ be the element of B such that $w_n^{(m)} = 0$ $(\forall\, n \in N\colon n \ne m)$ and $w_m^{(m)} = 1$. Since $(\forall\, n,m \in N\colon n \ne m)$ $1 = \|w^{(n)}\|_2$ and $\|w^{(n)} - w^{(m)}\|_2 = \sqrt{2}$, it is clear that $\{w^{(j)}\}_{j=1}^{\infty}$ has no $\mathfrak{T}_2$-convergent subsequence. This implies that B is not compact, that

(viii) $\ell_2(N)$, under $\mathfrak{T}_2$, is not semi-Montel.

However, the sequence $\{w^{(j)}\}_{j=1}^{\infty}$ evidently converges to 0 in the topology $\mathfrak{T}_\eta$. Hence,

(ix) $\mathfrak{T}_\eta$ is strictly coarser than $\mathfrak{T}_2$.

Thus, (vi) and (ix) imply

(x) $\ell_2(N)$, under $\mathfrak{T}_\eta$, is not a Mackey space.

Now, let $\{x^{(\alpha)}\}$ be any net in B . Then $(\forall\ n \in N)\ \sup_{\alpha} |x_n^{(\alpha)}| \leq 1$ and so there exists a subnet $\{x^{s(\beta)}\}$ of $\{x^{(\alpha)}\}$ and some $x \in K^N$ such that $(\forall\ n \in N)\ \lim_{\beta} |x_n^{s(\beta)} - x_n| = 0$. For each $n \in N$,

$$\left(\sum_{j=1}^{n} |x_j|^2 \right)^{\frac{1}{2}} = \lim_{\beta} \left(\sum_{j=1}^{n} |x_j^{s(\beta)}|^2 \right)^{\frac{1}{2}} \leq 1$$

and so x is in B . Let y be an arbitrary element of $\ell_2(N)$ and ε any positive number. Choose $m \in N$ such that

$$\sum_{j=m+1}^{\infty} |y_j|^2 < \left(\frac{\varepsilon}{4} \right)^2 .$$

Choose β_0 such that $(\forall\ \beta > \beta_0)$

$$\sup_{j=1}^{m} |x_j^{2(\beta)} - x_j| < \frac{\varepsilon}{2 \cdot \sum_{j=1}^{m} |y_j|} .$$

Then $(\forall\ \beta > \beta_0)$ the above and Hölder's Inequality imply

$$\|x^{s(\beta)} - x\|_y = \sum_{j=1}^{m} |x_j^{s(\beta)} - x_j| \cdot |y_j| + \sum_{j=m+1}^{\infty} |x_j^{s(\beta)} - x_j| \cdot |y_j| \leq$$

$$\left(\sup_{j=1}^{m} |x_j^{s(\beta)} - x_j|\right) \cdot \sum_{j=1}^{m} |y_j| + \|x_j^{s(\beta)} - x\|_2 \left(\sum_{j=m+1}^{\infty} |y_j|^2 \right)^{\frac{1}{2}} \leq$$

$$\frac{\varepsilon}{2} + 2 \left(\sum_{j=m+1}^{\infty} |y_j|^2 \right)^{\frac{1}{2}} \leq \frac{\varepsilon}{2} + 2 \cdot \left(\left(\frac{\varepsilon}{4} \right)^2 \right)^{\frac{1}{2}} = \varepsilon$$

This proves that $x^{s(\beta)}$ converges to x in the topology $\mathfrak{T}_\eta$. Consequently, by (vii),

(xi) $\ell_2(N)$, under $\mathfrak{T}_\eta$, is semi-Montel.

Let $\{x^{(\alpha)}\}$ be any $\mathfrak{J}_\eta$-Cauchy net in $\ell_2(N)$. Let the sequence $\{w^{(n)}\}$ be, as before, the characteristic functions of singletons. For each $n \in N$, we have

$$0 = \lim_{\alpha,\beta} \|x^{(\alpha)} - x^{(\beta)}\|_{w^{(n)}} = \lim_{\alpha,\beta} |x_n^{(\alpha)} - x_n^{(\beta)}| \ .$$

Thus, there exists some $x \in K^N$ such that $(\forall\ n \in N)\ \lim_\alpha x_n^{(\alpha)} = x_n$.

Assume $x \notin \ell_2(N)$. Choose an increasing sequence $\{s(n)\}_{n=1}^\infty$ in N such that $s(1) = 1$ and

$$r_n \equiv \sum_{j=s(n)}^{s(n+1)-1} |x_j|^2 \geq 1 \qquad\qquad (\forall\ n \in N) \ .$$

Let $y \in K^N$ be defined by $(\forall\ j \in N)$

(xii) $$y_j \equiv \frac{1}{n} \cdot |x_j| \cdot r_n^{-\frac{1}{2}} \quad \text{where} \quad s(n) \leq j \leq s(n+1)$$

Then

$$\sum_{j=1}^\infty |y_j|^2 = \sum_{n=1}^\infty \sum_{j=s(n)}^{s(n+1)-1} |y_j|^2 = \sum_{n=1}^\infty \frac{1}{n^2} < \infty$$

so that $y \in \ell_2(N)$. Since $x^{(\alpha)}$ is $\mathfrak{J}_\eta$-Cauchy, there exists $r \geq 0$ such that

$$r = \lim_\alpha \|x^{(\alpha)}\|_y = \lim_\alpha \sum_{n=1}^\infty |x_n^{(\alpha)} \cdot y_n| \ .$$

Thus, for each $m \in N$,

$$r \geq \overline{\lim_\alpha} \sum_{n=1}^m |x_n^{(\alpha)} \cdot y_n| = \sum_{n=1}^m |x_n \cdot y_n| \ .$$

This implies

$$r \geq \sum_{n=1}^\infty |x_n \cdot y_n|$$

but, by (xii) ,

$$\sum_{n=1}^{\infty} |x_n \cdot y_n| = \sum_{n=1}^{\infty} \sum_{j=s(n)}^{s(n+1)-1} |x_j \cdot y_j| = \sum_{n=1}^{\infty} \frac{1}{n} r_n^{\frac{1}{2}} \geq \sum_{n=1}^{\infty} \frac{1}{n} = \infty \quad ,$$

a contradiction. Hence, $x \in \ell_2(N)$.

Now let w be any element of $\ell_2(N)$ and ε any positive number. Since $\{x^{(\alpha)}\}$ is $\mathfrak{T}_\eta$-Cauchy, there exists α_0 such that $(\forall\ \alpha,\beta \geq \alpha_0)$

$$\frac{1}{3}\ \varepsilon > \|x^{(\alpha)} - x^{(\beta)}\|_w = \sum_{n=1}^{\infty} |x_n^{(\alpha)} - x_n^{(\beta)}| \cdot |w_n| \quad .$$

Pick $m \in N$ such that

$$\sum_{n=m+1}^{\infty} |x_n^{(\alpha_0)} - x_n| \cdot |w_n| < \frac{1}{3}\ \varepsilon \quad .$$

Choose α_1 such that $\alpha_1 > \alpha_0$ and $(\forall\ \alpha > \alpha_1)$

$$\sup_{n=1}^{m} |x_n^{(\alpha)} - x_n| < \frac{1}{3}\ \varepsilon / \sum_{n=1}^{m} |w_n| \quad .$$

Then $(\forall\ \alpha > \alpha_1)$

$$\|x^{(\alpha)} - x\|_w = \left(\sum_{n=1}^{m} + \sum_{n=m+1}^{\infty} \right) (|x_n^{(\alpha)} - x_n| \cdot |w_n|) \leq$$

$$\frac{\varepsilon}{3} + \sum_{n=m+1}^{\infty} |x_n^{(\alpha)} - x_n^{(\alpha_0)} + x_n^{(\alpha_0)} - x_n| \cdot |w_n| \leq$$

$$\frac{\varepsilon}{3} + \sum_{n=m+1}^{\infty} |x_n^{(\alpha)} - x_n^{(\alpha_0)}| \cdot |w_n| + \sum_{n=m+1}^{\infty} |x_n^{(\alpha_0)} - x_n| \cdot |w_n| <$$

$$\frac{\varepsilon}{3} + \frac{\varepsilon}{3} + \frac{\varepsilon}{3} = \varepsilon \quad .$$

It follows that $\lim_\alpha \|x^{(\alpha)} - x\|_w = 0$ and, in general, $x^{(\alpha)}$ converges to x in $\mathfrak{T}_\eta$. This proves

(xiii) $\ell_2(N)$, under $\mathfrak{T}_\eta$, is complete.

In view of (xiii), we know ([S] II.5.4)

(xiv) $\ell_2(N)$, under $\mathfrak{T}_\eta$, is the projective limit of Banach spaces.

E. The Space $\ell_1(N)$.

For each $x \in K^N$, let

(i) $$\|x\|_1 \equiv \sum_{n=1}^{\infty} x_n \quad .$$

Then, as is well known,

(ii) $\ell_1(N) \equiv \{x \in K^N: \|x\|_1 < \infty\}$ is a Banach space under the norm $\| \ \|_1$.

Write $\ell_\ell(N)$ for the set $\{x \in K^N: \lim_{n\to\infty} x_n$ exists and is finite$\}$. Then $\ell_\ell(N)$ is a linear subspace of K^N. For each $x \in \ell_\ell(N)$ define a function $f_x | \ell_1(N) \to K$ by

(iii) $$f_x(y) \equiv \sum_{n=1} x_n \cdot y_n \qquad (\forall\ y \in \ell_1(N)).$$

For all $x \in \ell_\ell(N)$ and $y \in \ell_1(N)$, we have

$$|f_x(y)| \le \sum_{n=1} |x_n \cdot y_n| \le (\sup_{n=1}^{\infty} |x_n|) \cdot \|y\|_1$$

so that

(iv) $\quad S \equiv \{f_x: x\in\ell_\ell(N)\}$ is a subspace of the dual $\ell_1(N)'$.

For each $n\in N$, let $w^{(n)}$ be the function on N such that $w_n^{(n)} = 1$ and $w_j^{(n)} = 0 \quad (\forall\ j\in N: j \neq n)$. Define the function $F_0|S \to K$ by

$$F_0(f_x) \equiv \lim_n x_n \qquad (\forall\ x\in\ell_\ell(N)).$$

Then F_0 is a linear functional on S and $(\forall\ x\in\ell_\ell(N))$

(v) $$\lim_n f_x(w^{(n)}) = F_0(f_x) .$$

We shall show

(vi) $$(\forall\ y\in\ell_1(N))\ (\exists\ x\in\ell_\ell(N)) \qquad f_x(y) \neq F_0(f_x) .$$

Let y be any element of $\ell_1(N)$. Choose $m\in N$ such that $\sum_{n=m}^{\infty} |y_n| < \frac{1}{2}$. Let $x\in\ell_\ell(N)$ be defined by $x_n \equiv 0 \quad (\forall\ n=1,2,\ldots,m-1)$ and $x_n = 1 \ (\forall\ n \geq m)$. Then

$$F_0(x) = 1 > \tfrac{1}{2} > \sum_{n=m}^{\infty} |y_n| \geq |\sum_{n=m} y_n| = |f_x(y)| .$$

This proves (vi) .

The sequence $\{w^{(n)}\}$ lies in the unit ball of $\ell_1(N)$. But (iv) , (v), and (vi) imply that no subsequence of $\{w^{(n)}\}$ can converge weakly to any $y\in\ell_1(N)$. This proves

(vii) $\quad \ell_1(N)$ is not semi-reflexive.

F. <u>The Space $c_{oo}(N)$</u> .

Let

(i) $$c_\infty(N) \equiv \{x\in K^N: (\exists\ m\in N)\ (\forall\ n>m)\ x_n = 0\} .$$

For each $x \in c_{oo}(N)$, let

(ii) $$\|x\|_u \equiv \max_{n=1}^{\infty} |x_n| \quad .$$

Then

(iii) $c_{oo}(N)$, under $\| \ \|_u$, is a normed linear space.

For each $n \in N$, let $v^{(n)}$ be defined by $v_j^{(n)} \equiv \frac{1}{j}$ $(\forall\ j=1,2,\ldots,n)$ and $v_j^{(n)} = 0$ $(\forall\ j > n)$. The sequence $\{v^{(n)}\}$ is clearly $\| \ \|_u$-Cauchy, and yet cannot have a limit in $c_{oo}(N)$. Since $\{v^{(n)}\}$ is in the unit ball of $c_{oo}(N)$,

(iv) $c_{oo}(N)$ is not quasi complete.

Let $A \equiv \{x \in c_{oo}(N) \colon |x_n| \leq \frac{1}{n} \quad (\forall\ n \in N)\}$.

Then A is convex, circled, closed, and absorbent, but not a neighborhood of zero in $c_{oo}(N)$. Thus,

(v) $c_{oo}(N)$ is not barreled.

Now we put a second topology T_s on $c_{oo}(N)$, the direct sum topology. We shall show

(vi) $c_{oo}(N)$, under T_s , is not metrizable.

Assume $\{B_n\}_{n=1}^{\infty}$ were a countable base for the neighborhood system of zero. Then $(\forall\ n \in N)$ $(\exists\ x^{(n)} \in\ ]0,\infty[^N)$ such that $B_n' \equiv \{y \in c_{oo}(N) \colon (\forall\ j \in N) |y_j| \leq x_j^{(n)}\}$ is a neighborhood of zero in $c_{oo}(N)$ and $B_n' \subset B_n$. Thus, $\{B_n'\}_{n=1}^{\infty}$ is a neighborhood base for zero. Define $x \in\ ]0,\infty[^N$ by letting $(\forall\ n \in N)$ $x_n = \frac{1}{2} x_n^{(n)}$. Then $B \equiv \{y \in c_{oo}(N) \colon (\forall\ j \in N)\ |y_j| < x_j\}$ is a T_s neighborhood of zero and

$(\forall\ n \in N)$ $B_n \not\subset B$: a contradiction. This proves (vi).

We put a third topology on $c_{oo}(N)$: the topology inherited from the product space K^N. The argument used in showing (iv) suffices to show that

(vii) $c_{oo}(N)$ is not quasi-complete; but the topology on $c_{oo}(N)$ is the projective topology from a complete Montel space K^N.

Since a locally convex direct sum of complete spaces is complete, then $c_{oo}(N)$ under T_s is complete. Hence T_2 is strictly finer than the topology on $c_{oo}(N)$ inherited from K^N. But the dual of $c_{oo}(N)$ with respect to either of the two topologies is the same (it can be identified with K^N, [S] p. 138). Thus

(viii) $c_{oo}(N)$, under the topology inherited from K^N, is not a Mackey space.

G. The Product Space K^N.

For each $n \in N$, define the semi-norm $\| \ \|_n$ on K^N by

(i) $$\|x\|_n \equiv |x_n| \qquad (\forall\ x \in K^N).$$

Then the topology generated by the family $\{\| \ \|_n\}_{n=1}^{\infty}$ of semi-norms is just the product topology. The metric d on K^N defined by

$$d(x,y) \equiv \sum_{n=1}^{\infty} \left(\frac{\|x-y\|_n}{1+\|x-y\|_n}\right)\left(\frac{1}{2^n}\right) \qquad (\forall\ x,y \in K^N)$$

gives the same topology and so

(ii) K^N is metrizable.

Since K is a complete Montel space,

(iii) $\quad K^N$ is a Frechet Montel space.

However, since a normed Montel space must be finite dimensional,

(iv) $\quad K^N$ is not a normed space.

H. An Inductive Limit (due to Komura). [YK]

Let $\mathcal{C}$ be the family of all countable infinite subsets of R and $(\forall A \in \mathcal{C})$ let

(i) $\quad E_A \equiv \{f \in K^R\colon (\forall\, x \in R\colon x \notin A) \quad f(x) = 0\}$

bear the topology $\mathfrak{T}_A$ inherited from the product K^R. Let

(ii) $\quad E \equiv \bigcup_{A \in \mathcal{C}} E_A$.

Let $\mathfrak{T}_1$ be the topology on E inherited from K^R and let $\mathfrak{T}_2$ be the topology on E where E is considered the locally convex inductive limit of the subspaces E_A, $A \in \mathcal{C}$. That

(iii) $\quad \mathfrak{T}_1 \subset \mathfrak{T}_2$

is evident. We shall prove the reverse inclusion.

For each subset X of E , let [X] be the $\mathfrak{T}_2$ closed, circled, convex hull of X . Let V be any $\mathfrak{T}_2$-neighborhood of zero in E . Then there exists a family $\{V_A\colon A \in \mathcal{C}\}$ such that

(iv) $\quad [\bigcup_{A \in \mathcal{C}} V_A] \subset V$

and $(\forall\, A \in \mathcal{C})$ V_A is a neighborhood of zero in E_A . For each $A \in \mathcal{C}$, choose a finite subset F_A of A and $\varepsilon_A > 0$ such that, if

$V'_A \equiv \{f \in E_A\colon |f(x)| < \varepsilon_A \quad (\forall\, x \in F_A)\}$, then

(v) $\qquad V'_A \subset V_A$.

Let $(\forall\ A\in\mathcal{C})$ I_A be the set $\{x\in A:\ x\notin F_A\}$, and let $B \equiv \{x\in R:\ x\notin \bigcup_{A\in\mathcal{C}} I_A\}$. If B were infinite, then $(\exists\ A\in\mathcal{C})$ $A \subset B$; then $I_A \cap B \neq \Phi$, which is impossible. Hence, B is finite, and $(\exists\ D\in\mathcal{C})$ $B \subset D$. Let

(vi) $\qquad w \equiv \{f\in E:\ (\forall\ x\in B)\ |f(x)| < \varepsilon_D\}$.

We proceed to show

(vii) $\qquad W \subset V$.

Let f be an arbitrary element of W . Then $(\exists\ C\in\mathcal{C})$

(viii) $\qquad D \subset C$ and $f \in E_C$.

Let $\{x_n\}_{n=1}^m$ be the set of all elements in $F_D \cup F_C$ but not in B . For each $n = 1,2,\ldots,m$, choose $A_n \in \mathcal{C}$ such that $x_n \in I_{A_n}$. Let

$$\{y_j\}_{j=1}^p \equiv F_C \cap F_D \cap \left(\bigcap_{n=1}^m A_n\right)$$

and

$$\{z_j\}_{j=1}^q \equiv \{x\in F_C \cup F_D \cup \left(\bigcup_{n=1}^m F_{A_n}\right) :\ x\notin \{y_j\}_{j=1}^p\} .$$

Let

$$a_1 \equiv \frac{1}{\varepsilon_C} \max_{j=1}^{p} |f(y_j)|$$

and $(\forall\ j=2,3,\ldots,q)$

(ix) $\qquad a_j \equiv \dfrac{1-a_1}{q-1}$ so that $\displaystyle\sum_{j=1}^{q} a_j = 1$.

For each $j=1,2,\ldots,q$ define $h_j \in E$ by

$$h_j(x) \equiv \begin{cases} \frac{1}{a_1} f(y_k) & \text{if } j=1 \text{ and } (\exists\, k=1,2,\ldots,p)\ x=y_k, \\ \frac{1}{a_j} f(z_j) & \text{if } x = z_j , \\ 0 & \text{otherwise} . \end{cases}$$

For each $j=1,x,\ldots,q$ choose $G_j \in \mathcal{C}$ from $\{C,D\} \cup \{A_n: n=1,\ldots,m\}$ such that $z_j \notin F_{G_j}$. Then $(\forall\, j=1,2,\ldots,q)$ $h_j \in V_{G_j}$ and

$f = \sum_{j=1}^{q} a_j h_j$. From (ix) it follows that

$$f \in [V'_C \cup V'_D \cup (\bigcup_{n=1}^{m} V_{A_n})]$$

which, by (iv) and (v), proves (vii). Thus $\mathfrak{I}_2 \subset \mathfrak{I}_1$ and so, by (iii),

(x) $\quad \mathfrak{I}_1 = \mathfrak{I}_2$.

Note that $(\forall\, A \in \mathcal{C})$ E_A is topologically isomorphic with the product K^A; thus,

(xi) $\quad E_A$ is a complete Montel space.

Let S be the set $\{f \in E: (\forall\, x \in R)\ |f(x)| \leq 1$. Then S is a closed and bounded subset of E. For each $A \in \mathcal{C}$, let f_A be the function in E_A such that $f_A(x) = 1$ $(\forall\, x \in A)$. If the elements of $\mathcal{C}$ are ordered by inclusion, then $\{f_A\}_{A \in \mathcal{C}}$ is a net in S which converges to the constant function 1 in K^R. It follows that S is not $\sigma(E,E')$-compact. This means

(xii) E is not quasi-complete.

I. <u>The Space $c_{oo}(R)$</u>.

Let $c_{oo}(R)$ be the set of all functions in K^R with finite support. Let $\mathfrak{S}$ be the set of all sequences in R and, for each $s \in \mathfrak{S}$, let

(i) $$A(s) = \{x \; c_{oo}(R): |x_{s(n)}| \le \frac{1}{n} \; (\forall \; n \in N)\} \; .$$

The family $\{A(s): s \in \mathfrak{S}\}$ constitutes a base for a locally convex topology T on $c_{oo}(R)$.

Let

$$B = \{x \in c_{oo}(R): \sum_{a \in R} |x_a| \le 1\} \; .$$

It is evident that B is a barrel. For each $s \in S$, there exist some $x \in A(s)$ and some $a \in R$ such that $|x_a| > 1$. Thus, B contains no $A(s)$ as a subset and, consequently, is not a neighborhood of 0 . This proves

(ii) $c_{oo}(R)$, under T , is not barreled.

We now show that

(iii) $c_{oo}(R)$, under T , is a projective limit of Montel spaces.

Let $\mathfrak{Z}$ be the family of all countable subsets of R and write $W < V$ iff $W \subset V$ for all $V, W \in \mathfrak{Z}$. Let $(\forall \; W \in \mathfrak{Z})$

(iv) $$E_W = \{x \in c_{oo}(R): x_a = 0 \; (\forall \; a \notin W)\} \; .$$

It can be seen that E_W, under the topology relativized from T , is isomorphic to a countable direct sum of the scalars; as such

(v) E_W is a Montel space.

For each $W, V \in F$ such that $W < V$, define $f_{VW} | E_V \to E_W$ for all $x \in E_V$,

(vi) $$f_{VW}(x)_a = \begin{cases} x_a & \text{if } a \in W \\ 0 & \text{if } a \notin W \end{cases} .$$

Then each f_{VW} is continuous and we may consider the projective limit

(vii) $$E = \lim_{\to} f_{VW}(E_V) .$$

For each $g \in E$, there is evidently exactly one $\psi(g) \in c_{oo}(R)$ such that

(viii) $$\psi(g)_a = g_V(a) \quad \text{for all } V \in F \text{ and } a \in V .$$

That the map $\psi | E \to c_{oo}(R)$ thus defined is a surjective topological isomorphism, is easy to check. This proves (iii).

APPENDIX

A. An Inductive Limit of Locally Convex Subspaces which bears the Indiscrete Topology (a slight modification of Komura's Example 1 in [YK]).

Let A be the set $\{ \frac{m}{2^n}: n \in N , m = 0,1,\dots,2^n \}$ and let $I \equiv \{x \in [0,1]: x \notin A\}$. Let $\mathfrak{F}$ be the family of all non-void open sub-intervals of $[0,1]$ with endpoints in A . For $S \in \mathfrak{F}$, write $\lambda(S)$ for the length of S . Let $\mathfrak{G}$ be the collection of all countable sub-families $\mathfrak{H}$ of $\mathfrak{F}$ such that

(i) $$(\forall\ S,M \in \mathfrak{H}:\ S \neq M) \quad S \cap M = \Phi \text{ and } \sum_{S \in \mathfrak{H}} \lambda(S) = 1 \quad .$$

We assume functions which agree except at points in A are identified and for each $\mathfrak{H} \in \mathfrak{G}$, let

(ii) $$E_{\mathfrak{H}} \equiv \{f \in K^I: (\forall\ S \in \mathfrak{H})\ f \text{ is constant on } I \cap S \quad .$$

For each $\mathfrak{H} \in \mathfrak{G}$ and $S \in \mathfrak{H}$, define the semi-norm $\| \|_S^{\mathfrak{H}}$ on $E_{\mathfrak{H}}$ by $(\forall\ f \in E_{\mathfrak{H}})$

(iii) $$\|f\|_S^{\mathfrak{H}} \equiv |f(x)| \cdot \lambda(S) \qquad \text{for any } x \in I \cap S \ .$$

If $(\forall\ \mathfrak{H} \in \mathfrak{G})$ $T_{\mathfrak{H}}$ is the topology on $E_{\mathfrak{H}}$ induced by the family $\{\| \|_S^{\mathfrak{H}}: S \in \mathfrak{H} \}$ of semi-norms, then

(iv) $E_{\mathfrak{H}}$, under $T_{\mathfrak{H}}$, is a locally convex space.

For $\mathfrak{H} \in \mathfrak{G}$, $E_{\mathfrak{H}}$ is a metrizable space and a metric $d^{\mathfrak{H}}$ yielding the topology $T_{\mathfrak{H}}$ may be defined as follows: $(\forall\ f,g \in E_{\mathfrak{H}})$

(v) $$d^{\mathcal{H}}(f,g) \equiv \sum_{S\in\mathcal{H}} \lambda(S)\cdot \frac{|f(x_S)-g(x_S)|}{|f(x_S)-g(x_S)|+1}$$

where $(\forall S\in\mathcal{H})\, x_S\in S$. It is clear that $(\forall \mathcal{H}\in\mathcal{G})$

(vi) $$d^{\mathcal{H}}(f,g) = \int_0^1 \frac{|f(x)-g(x)|}{|f(x)-g(x)|+1}\, dx$$

for all f and g in $E_{\mathcal{H}}$.

For $\mathcal{H}$ and $\mathcal{J}$ in $\mathcal{G}$, write

(vii) $\mathcal{J} < \mathcal{H}$ if $(\forall T\in\mathcal{H})(\exists S\in\mathcal{J})\ T\subset S$.

It is evident that, if $\mathcal{J}<\mathcal{H}$, then $E_{\mathcal{J}}$ is a subspace of $E_{\mathcal{H}}$; furthermore, by (vi), the topology $T_{\mathcal{J}}$ is just the topology $T_{\mathcal{H}}$ relativized to $E_{\mathcal{J}}$.

For $\mathcal{H}$ and $\mathcal{J}$ in $\mathcal{G}$, write $\mathcal{H}\wedge\mathcal{J}$ for the family $\{T\cap S\colon T\in\mathcal{H},\ S\in\mathcal{J}\}$; then $\mathcal{H}\wedge\mathcal{J}$ is in $\mathcal{G}$ and both $\mathcal{H}\wedge\mathcal{J}>\mathcal{H}$ and $\mathcal{H}\wedge\mathcal{J}>\mathcal{J}$. Thus, if we define

(viii) $$E \equiv \bigcup_{\mathcal{H}\in\mathcal{G}} E_{\mathcal{H}} ,$$

then E may be regarded as the inductive limit of the family $\{E_{\mathcal{H}}\colon \mathcal{H}\in\mathcal{G}\}$ of subspaces, the topology $\mathfrak{T}$ on E being the finest locally convex topology of which the restriction to each $E_{\mathcal{H}}$ is coarser than $T_{\mathcal{H}}$.

We shall show

(ix) $T = \{E,\Phi\}$, the indiscrete topology on E

or, what is equivalent,

(x) there exists no non-zero continuous linear functional F on E .

Let F be a continuous linear functional on E . For each $S\in\mathfrak{F}$, write f_S for the function in E such that $f(x) = 1\ (\forall x\in S\cap I)$ and

$f(x) = 0$ for all other $x \in I$. Suppose that $\mathcal{H}$ is in $\mathcal{G}$ and $\{S \in \mathcal{H}: F(f_S) \neq 0\}$ is infinite. Then for proper α_S, $F(\alpha_S f_S) = 1$. Define $f = \sum_{S \in \mathcal{H}} \alpha_S f_S \in E_{\mathcal{H}}$. Since the net of finite sums $\sum_{\text{finite}} \alpha_S f_S$ converges to f in $E_{\mathcal{H}}$ and F is continuous on $E_{\mathcal{H}}$, $F(f) = \lim F(\sum_{\text{finite}} \alpha_s f_s) = \lim \sum_{\text{finite}} F(\alpha_S f_S)$ which does not exist. Hence

(xi) $\{S \in \mathcal{H}: F(f_S) \neq 0\}$ is a finite set.

Assume

(xii) $(\exists\, x \in [0,1])\ (\forall\, \varepsilon > 0)\ (\exists\, S_\varepsilon \in \mathfrak{F})\ S_\varepsilon \subset]x-\varepsilon, x+\varepsilon[$ and $F(f_S) \neq 0$.

If there exists a decreasing sequence $\{s(n)\}_{n=1}^{\infty}$ tending to 0 for which $(\forall\, n \in N)$ x is not in the closure of $S_{s(n)}$, then the sequence can be taken such that $(\forall\, n \in N)\ [x-s(n+1), x+s(n+1)] \cap S_{s(n)} = \Phi$ in this case, it is clear that $(\forall\, n,m \in N: n \neq m)\ S_{s(n)} \cap S_{s(m)} = \Phi$ and there exists some $\mathcal{H} \in \mathcal{G}$ such that $(\forall\, n \in N)\ S_{s(n)} \in \mathcal{H}$. But $(\forall\, n \in N)\ F(f_{S_{s(n)}}) \neq 0$, which violates (xi). Thus

(xiii) $(\exists\, \varepsilon > 0)\ (\forall\, 0 < \delta < \varepsilon)$ x is in the closure of S_δ

Choose $\delta(1)$ less than ε and $\mathcal{H} \in \mathcal{G}$ such that $S_{\delta(1)} \in \mathcal{H}$. Choose $\mathcal{J} \in \mathcal{G}$ such that $\mathcal{H} < \mathcal{J}$ and x is in the closure of no element of $\mathcal{J}$. If $\{B(n)\}_{n=1}^{\infty}$ is an enumeration of those elements of $\mathcal{J}$ which are subsets of $S_{\delta(1)}$, then (vi) implies $f_{S_{\delta(1)}} = \lim_{m=1}^{\infty} \sum_{n=1}^{m} f_{B(n)}$ so that $F(f_{S_{\delta(1)}}) = \lim_{m=1}^{\infty} \sum_{n=1}^{m} F(f_{B(n)})$. Thus, (xii) implies $(\exists\, m \in N)$

$F(f_{B(m)}) \neq 0$. Let $D(1) \equiv B(m)$. Since x is not in the closure of $D(1)$, we can find $\delta(2) < \delta(1)$ such that $S_{\delta(2)} \cap D(1) = \Phi$ and repeat the above argument to obtain $D(2) \in \mathfrak{F}$ such that $D(2) \subset S_{\delta(2)}, F(f_{D(2)}) \neq 0$, and x is not in the closure of $D(2)$. In general, we can construct a decreasing sequence $\{\delta(n)\}_{n=1}^{\infty}$ of positive numbers and a sequence $\{D(n)\}_{n=1}^{\infty}$ in $\mathfrak{F}$ such that $(\forall\ n \in N)\ D(n) \subset S_{\delta(n)}$ and $D(n) \cap [x-\delta(n+1), x+\delta(n+1)] = \Phi$. Evidently there exists some $\mathcal{J} \in \mathcal{G}$ such that $\{D(n) : n \in N\} \subset \mathcal{J}$. But this contradicts (xi). It follows that assumption (xii) is not true, that

(xiv) $\quad (\forall\ x \in [0,1])\ (\exists\ \varepsilon_x > 0)\ (\forall\ S \in \mathfrak{F}: S \subset]x-\varepsilon_x, x+\varepsilon_x[)\ F(f_S) = 0$.

Since $[0,1]$ is compact, $(\exists\ \varepsilon > 0)\ (\forall\ y \in [0,1])\ (\exists\ x \in [0,1])$ $]y-\varepsilon, y+\varepsilon[\subset]x-\varepsilon_x, x+\varepsilon_x[$. It follows that $(\forall\ S \in \mathfrak{F} : \lambda(S) < \varepsilon)\ F(f_S) = 0$. Let M be any element of $\mathfrak{F}$. Then choose $\mathcal{H} \in \mathcal{G}$ such that $M \in \mathcal{H}$ and choose $\mathcal{J} \in \mathcal{G}$ such that $\mathcal{H} < \mathcal{J}$ and $(\forall\ S \in \mathcal{J} : S \subset M)\ \lambda(S) < \varepsilon$. If $\{S_n\}_{n=1}^{\infty}$ is an enumeration of the set $\{S \in \mathcal{J} : S \subset M$, then (vi) implies that $\lim_m \sum_{n=1}^{m} f_{S_n} = f_M$ in $E_{\mathcal{J}}$. Then

$$F(f_M) = \lim_m \sum_{n=1}^{m} F(f_{S_n}) = 0 .$$

This proves

(xv) $\quad F(f_M) = 0 \qquad (\forall\ M \in \mathfrak{F})$.

If f is an arbitrary function in E , then $(\exists\ \mathcal{H} \in \mathcal{G})\ f \in E_{\mathcal{H}}$. If $\{S_n\}_{n=1}^{\infty}$ is an enumeration on $\mathcal{H}$ and if $(\forall\ n \in N)\ x_n \in S_n \cap I$, then (v) implies

$$f = \lim_m \sum_{n=1}^{m} f(x_n) \cdot f_{S_n} .$$

From (xv), it follows that $F(f) = 0$. This proves (x).

B. The Cardinality of a Linear Space.

We shall prove for use in Example (V.B) the following:

(i) if V is an infinite dimensional linear space over K with a Hamel basis B, then $\overline{\overline{V}} = \overline{\overline{B}}$;

(ii) if $\mathfrak{B}$ is the family of subsets of an infinite set B, and if $c_{oo}(\mathfrak{B})$ is the family $\{f \in K^{\mathfrak{B}}: f$ has a finite support$\}$, then $\overline{\overline{c_{oo}(\mathfrak{B})}} = \overline{\overline{K}}^{B} = \overline{\overline{\mathfrak{B}}}$.

That $\overline{\overline{B}} \leq \overline{\overline{V}}$ is trivial. As is well-known, the family $\mathfrak{F}(B)$ of finite subsets of B has the cardinality $\overline{\overline{B}}$. Since for each $S \in \mathfrak{F}(B)$, $\overline{\overline{\text{span } S}} = \overline{\overline{K}}$ and $V = \bigcup_{S \in \mathfrak{F}(B)} \text{span } S$, it follows that

$$\overline{\overline{V}} \leq \overline{\overline{K}} \cdot \overline{\overline{F(B)}} = \overline{\overline{\mathfrak{F}(B)}} = \overline{\overline{B}} .$$

This proves (i).

In (ii), that $\overline{\overline{c_{oo}(\mathfrak{B})}} = \overline{\overline{\mathfrak{B}}}$ follows from (i), since characteristic functions on elements of $\mathfrak{B}$ form a basis for $c_{oo}(\mathfrak{B})$. But, as is well-known, $\overline{\overline{K}}^{B} = \overline{\overline{\{0,1\}}}^{B} = \mathfrak{B}$. This proves (ii).

REFERENCES

1. Hewitt, E., Stromberg, K.; 1965, Real and Abstract Analysis, Springer-Verlag. [HS]

2. Horvath, J.; 1966, Topological Vector Spaces and Distributions, Vol. 1, Addison-Wesley. [H]

3. Komura, Y.; 1964, Math. Ann., Vol. 153, 150-162. [YK]

4. Köthe, G.; 1960, Topologische Lineare Räume, Springer-Verlag. [K]

5. Robertson, A., Robertson, W.; 1964, Topological Vector Spaces, Cambridge University Press. [R]

6. Schaefer, H.; 1966, Topological Vector Spaces, Macmillan. [S]

7. Yoshida, K.; 1965, Functional Analysis, Springer-Verlag. [Y]